Anatomy
of the dog

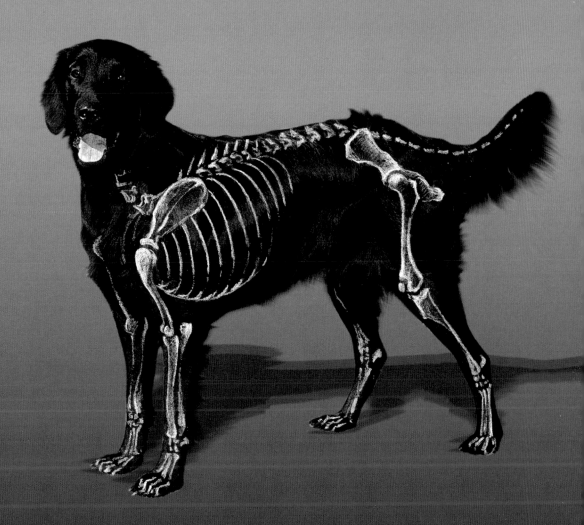

Anatomy of the dog

In straight forward terms

by Kerstin Mielke

CADMOS

Impressum
Copyright © 2007 by Cadmos Verlag GmbH, Schwarzenbek
Copyright of this edition © 2010 by Cadmos Books, Great Britain
Translated by Ute Weyer MRCVS
Title Photograph: Lehari/Mähler
Design by Ravenstein & Partner, Verden
Photographs: Dr Gabiele Lehari, Kerstin Mielke
Graphics: Maria Mähler, Kerstin Mielke
Editorial: Dr Gabriele Lehari, Dr Sarah Binns MRCVS,
Christopher Long
Printed by: Westermann Druck, Zwickau

British Library Cataloguing in Publication Data
A catalogue record of this book is available from the British Library.

Printed in Germany
ISBN 978-3-86127-979-2

Contents

Foreword 7

Introduction
to anatomy 9

The locomotor system 13

The bones 14
The joints 23
    Selected joints and
    their functions 28
The skeletal muscles 30
    How does a muscle contract? 33
    Muscles important for movement 35
    How does the dog move? 44
    Are coordinated movements
    inherited? 47
    Where does the energy for
    muscular activity come from? 49
Other muscles 51
    Respiratory muscles 51
    Abdominal muscles 51
    Muscles of the head 52
    Skin muscles 53
    Muscles of the tail 53

Internal organs 54

Respiration 55
Digestion and excretion 57
Reproduction 60

Heart and circulation 63

The heart 63
Circulation 66

The lymphatic system 68

Skin and hair 71

The nervous system 75

Central nervous system 76
Autonomic nervous system 77
Peripheral nervous system 77

The senses 79

Smell 80
Hearing 81
Vision 83
Touch 85
Taste 87

Appendix 88

English terms of location 89
Latin terms of location 89
The skeleton – English names 90
The skeleton – Latin names 91
The muscles – English names 92
The muscles – Latin names 93

The organs – English names 94
The organs – medical names 95
Important nerves and functions
of the relevant muscles 96

Literature 96

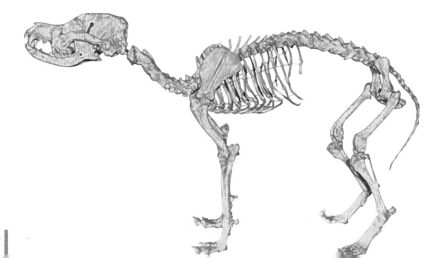

(Photo: Lehari)

Foreword

The dog belongs to the zoological class of vertebrates and shows many anatomical similarities to other mammals, including humans. These similarities are most pronounced among the carnivores.

Being a carnivore and therefore a predator the dog has a strong, well-muscled body plus a heart and lungs capable of extreme endurance. Sharp teeth enable the dog to tear apart its prey. Some modern breeds have developed inferior qualities to a certain degree, for instance breathing problems caused by the breeding of dogs with extremely short noses, or smaller body sizes that result in more fragile backs or legs. Despite this, or maybe even because of it, dogs are a fascinating species and their body shape and the harmony of their movements enthral humans time and again. This is not surprising – dogs are excellent runners! Therefore, the chapter about the locomotor system is especially detailed.

The body shape and harmonious movements of a dog are always fascinating. (Photo: Lehari)

This book sets out to explain in a concise manner many interesting aspects of the anatomy of your dog. Where it is helpful some physiological facts have been added. (Physiology explains the functions of the body.)

I have used mainly English terms and descriptions in this book so that it can be understood easily by a lay person. The appendix lists the technical terms for those who wish to familiarise themselves with them. The inserted boxes contain interesting facts and tips for everyday use.

I hope you will find this book interesting and that you share my enthusiasm for the anatomical side of dogs!

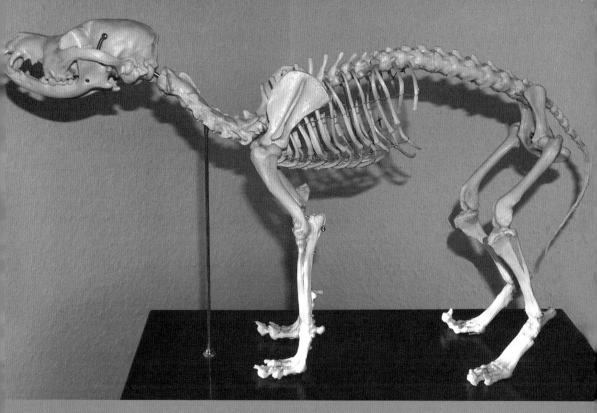

(Photo: Mielke)

Introduction
to anatomy

The term 'anatomy' stems from the Greek word 'anatome' which means 'dissect' or 'cut apart'. Nowadays we describe anatomy as everything to do with the structure of the body. Scientifically, anatomy is divided into many categories, such as general, comparative, macroscopic and systemic anatomy.

The different anatomical locations within the body are generaly described using Latin terms. Most people will have come across such terms when they are used by their vet or physiotherapist – soon you will know what they mean!

Should you have some knowledge of Latin from your school days do not be surprised when some names, e.g. of muscles, cannot be translated literally. It is correct that the musculus triceps brachii has four branches and a musculus extensor is also a flexor. The reason for this is that all the names stem from the same anatomical nomenclature. The nomenclature serves as the basis for comparative anatomy, which ensures that the same parts of the mammalian body in different species are always given the same name. However, depending on their actual function within each species, the body parts may have quite a different shape.

Introduction to anatomy basic terms. (Photo: Mielke)

Adduction and abduction of the hip joint. (Photo: Mielke)

Extension and flexion of the stifle joint. (Photo: Lehari)

We distinguish between various different main body parts or locations. Many names of the main body parts, such as head, chest, abdomen or legs, are also used for the human body.

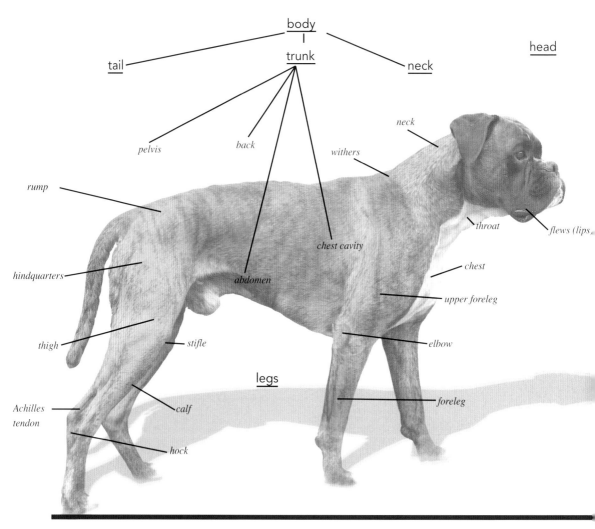

body
|
trunk

tail

neck

head

neck

pelvis

back

withers

rump

chest cavity

throat

flews (lips

hindquarters

abdomen

chest

upper foreleg

thigh

stifle

elbow

legs

Achilles
tendon

calf

foreleg

hock

Parts of the dog's body. (Photo: Mielke)

The locomotor system

The locomotor system of a dog is divided into two parts: active and passive. The passive part includes bones and joints whereas the skeletal muscles form the active part of the system.

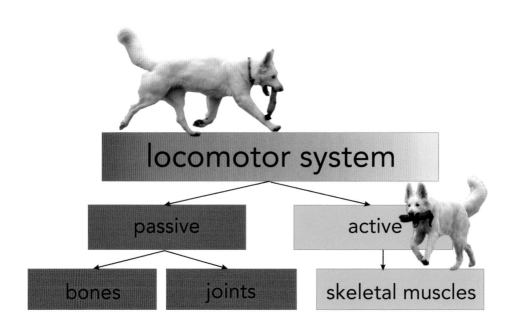

The locomotor system of a dog is divided into an active and a passive part.
(photos and graphics: Mielke)

The bones

The various bones in the body are connected to each other by joints (more details about joints are given in the next section). All bones are present as pairs and are located in the same places on each side of the body, with the exception of the vertebrae, the bones of the skull and the breastbone.

> Dogs have around 300 bones. The exact number differs depending on the breed and individual.

Bones consist of one third organic material (collagen fibres) and two thirds inorganic material. Collagen fibres are a type of connective tissue and they are located wherever stability is needed. If these fibres are boiled, the protein in the fibres turns to a glue-like substance. The Greek word 'colla' translates to glue.

Calcium phosphate makes up the majority of the inorganic material. After tooth enamel, bones are the hardest structure of the body. Their strength is due not only to the high level of mineralisation but also to the structure of the bones. The technical term is called a trajectory structure. The bone lamellae align along (invisible) lines within the bone providing maximal resistance to pulling and pushing forces. All bony structures in the body are covered by a delicate outer layer called the periosteum.

Bones undergo constant remodelling. Tissue is being broken down and rebuilt continuously. As long as the exercise regime and nutrition are sufficient the breakdown and rebuilding rates are precisely balanced.

On the basis of their shape there are four groups of bones: long bones, short bones, flat bones and sesamoid bones.

• **Long bones:** these are the large bones that make up the extremities, i.e. the front and hind legs.

Long bones consist of a shaft and two end pieces. Inside the shaft is a cavity that contains the bone marrow. Bone marrow plays an essential part in the production of blood cells, particularly in juvenile animals. The ends of the bone are made of compact bone filled with sponge-like bone lamellae. The ends are separated from the shaft at birth by a layer of cartilage, and they form the growth plates. When the animal is fully grown the growth plates close and the cartilage ossifies, thus turning into solid bone. Although the shape of the bone is fully developed at birth the bone structure is not. There are many reasons for this. For example, if

the pregnant bitch had to carry numerous pups with fully developed bones the weight would be significantly greater and the massively increased calcium and phosphorus requirements could not be met by the mother. Cartilage is also more flexible, which makes the birthing process easier and reduces the risk of injury to the newborn pups.

The long bones ossify from the outside inwards. This leads to the formation of a bony 'sleeve' surrounding the shaft. The bone thickens by adding bone cells on the outside. In order to avoid

The main components of a long bone.
(Figure: Mielke)

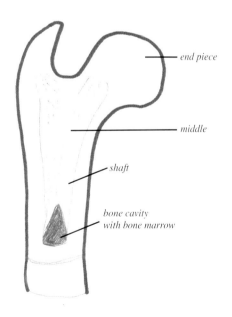

end piece

middle

shaft

bone cavity
with bone marrow

the bone becoming too thick the inner bone cells are broken down, thus enlarging the bone marrow cavity.

Shortening the long bones of the forearm (radius and ulna) and enlargement of the joints has led to the short-legged appearance of the Dachschund (dwarfism). Some other small dogs, however, have bones that are shortened without any deformation, e.g. Toy Poodles have all their body parts in the correct proportion to each other without any sign of dwarfism.

• **Short bones** are, as the name implies, relatively small bones, e.g. vertebrae or ankle bones. Their shape can vary significantly from round to rectangular or curved. They consist of a narrow bone cortex filled with a spongy bone structure.

• **Flat bones** are compact bone plates, e.g. the shoulder blades and pelvis, but the ribs and numerous bones in the skull also belong to this category.

• **Sesamoid bones** are special bones attached to or supporting tendons. One ex-

The main components of the skeleton. The number of vertebrae is included in the pictorial representation of the relevant part of the spine. (Photo: Mielke)

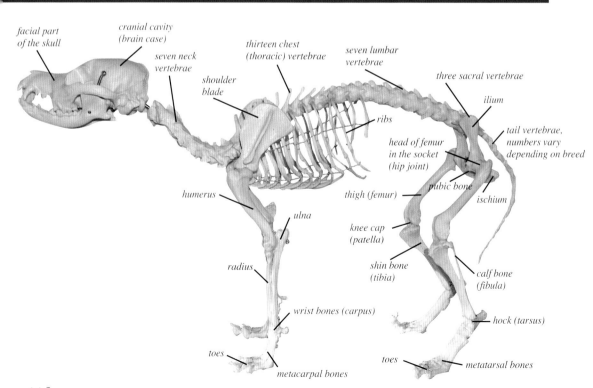

facial part of the skull

cranial cavity (brain case)

seven neck vertebrae

thirteen chest (thoracic) vertebrae

seven lumbar vertebrae

three sacral vertebrae

shoulder blade

ilium

ribs

tail vertebrae, numbers vary depending on breed

head of femur in the socket (hip joint)

humerus

thigh (femur)

pubic bone

ischium

ulna

knee cap (patella)

radius

shin bone (tibia)

calf bone (fibula)

wrist bones (carpus)

hock (tarsus)

toes

toes

metatarsal bones

metacarpal bones

List of palpable bone parts

Shoulder blade	anterior edge
	spinal edge (the cartilage that forms the withers can be felt)
	posterior rim
	scapular spine
	protrusion of scapular spine
Upper foreleg (humerus)	large tubercle
	groove
	lateral tubercle
Radius	head of radius
	small spur (middle)
Ulna	ulnar tubercle
	small spur (lateral)
Wrist (carpus)	sesamoid bone
Metacarpal bone	all bones of the front foot (five toes)
Pelvis	rim of ilium
	tubercle of ischium
	sacrum
Upper hind leg (femur)	trochanter
	bony protrusion (lateral and middle)
Knee cap (patella)	difficult to find, very small and embedded as a sesamoid bone in the tendon of the upper thigh muscle
Shin bone (tibia)	crest of the shin bone
	joint condyle (middle)
	hock (middle)
Calf bone (fibula)	small head near the condyle
	hock (lateral)
Hock	calcaneus
	tarsus
Metatarsal bone	all bones of the hind foot (as in the front leg but only four instead of five toes)
Toes	all toe bones including the claw bone

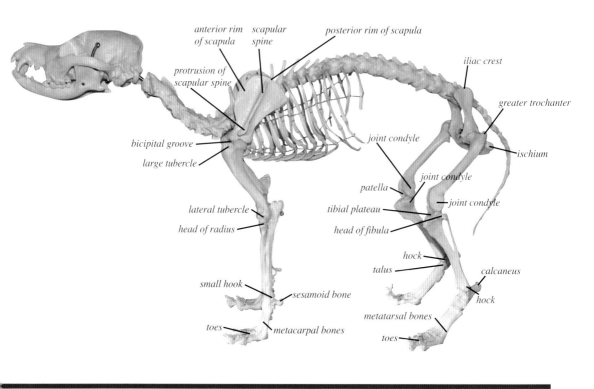

Palpable bones of the skeleton. (Photo: Mielke)

ample is the knee cap, which is attached to the tendon of the large thigh muscle. There is more about this in the section referring to joints.

All the bones together are called the skeleton; cartilage and connective tissue complete it. Apart from its function in locomotion, the skeleton protects important organs such as the brain and spinal cord. Many significant bony landmarks can be felt through the skin; all you need is a little bit of practice, some patience and basic anatomical knowledge. You can try

it for yourself! Often these palpable landmarks are areas of attachment of ligaments or muscles and therefore of special interest.

When you study the List – or even better your dog – you will realise that dogs have five toes in front but only four on their hind legs. The fifth toe is often present in the shape of a dewclaw. A dewclaw is defined as a claw without any inner bony parts; only the actual nail is developed. Many dogs do not have one, but for some breeds it is a typical attribute (e.g.

Briards). Unfortunately it is relatively common for dogs to injure their dewclaws or even tear them off completely. It is possible to have them removed by a vet, but whilst dewclaw removal is still legal in the UK it is only recommended for medical reasons.

The shoulder of the dog consists of the shoulder blade, the collar bone (clavicle) and the coracoid process. The collar bone and coracoid process are only rudiments because the foreleg has no grip function in the dog, unlike humans. The collar bone is merely a tendinous line in the leg-head muscle (brachiocephalicus).

The foreleg consists of the shoulder blade (scapula), upper forearm (humerus), radius, ulna, knee (patella), foot and toes.

The pelvis is comprised of three bones arranged in a ring: the ilium, which is formed in the shape of a wing, rising on each side of the pelvis; the ischium, which forms the middle portion of the pelvis; and the pubic bone, the bone at the base of the pelvic structure. The three bones are not fused in young puppies but once the dog is fully grown the different parts join completely.

The hind legs comprise the thigh (femur), knee cap (patella), shin bone (tibia), calf bone (fibula), hock (tarsus), foot and toes.

The chest or thorax is shaped by the ribs and spine.

The spine is made up of seven neck vertebrae, thirteen chest vertebrae, seven lumbar vertebrae, three sacral vertebrae

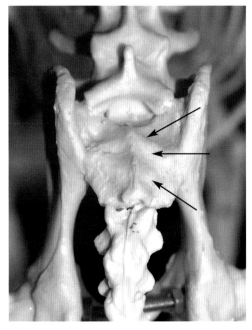

Once a dog is fully grown the three vertebrae of the sacrum fuse. This area is marked with the black arrows.

The pelvis as seen from one side (Photos: Mielke)

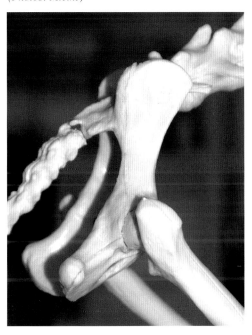

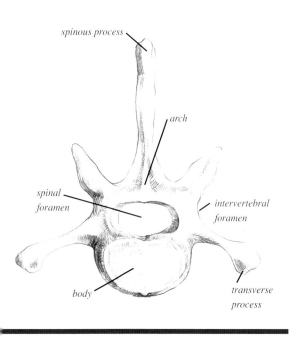

spinous process

arch

spinal foramen

intervertebral foramen

body

transverse process

Structure of a vertebra.
(Graphic: Maehler)

gel-like soft material surrounded by an outer shell of fibrous tissue.

A disc can tear, often owing to old age as the discs tend to calcify over the years. This calcification can lead to disc material protruding into the spinal canal, which results in compression of a nerve bundle and causes severe pain or even paralysis. This condition is common in small dogs with long backs such as the Dachschund and Bassett. Other breeds, however, are also susceptible.

The following bone parts can be palpated easily: the lateral aspects of the atlas (first neck vertebra), the spinal processes and depending on the breed a varying number of tail vertebrae (usually 20–23). The spine holds and protects the spinal cord and is also essential for posture and movement. Without the spine a dog could not support itself and therefore could neither stand nor walk.

Each vertebra is a strong, cylindrical bone with a round hole in its centre where the spinal cord lies. It has small openings along its sides that allow the spinal nerves to pass through.

Between the vertebrae lie the spinal discs. They function as shock absorbers and cushions to alleviate impact. Only the first and second vertebrae are not separated by a disc. The discs consist of a

Structure of a spinal disc.
(Graphic: Maehler)

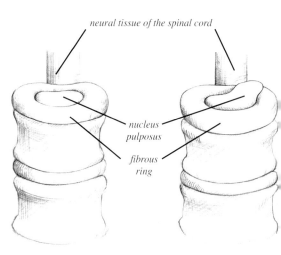

neural tissue of the spinal cord

nucleus pulposus

fibrous ring

healthy spinal disc

prolapsed spinal disc – protrudes into the spinal canal and compresses spinal nerves

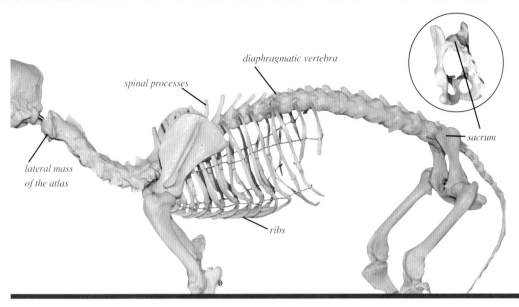

diaphragmatic vertebra

spinal processes

sacrum

*lateral mass
of the atlas*

ribs

Palpable bones of the spine. The sacrum is located behind the ilium when looking at the skeleton
from the side; therefore it is shown in the insert.
(Photo: Mielke)

The ribs are classified as follows:

1st to 9th ribs	true ribs – attached directly to the breastbone, supporting the chest cavity
10th to 12th ribs	false ribs – attached to the breastbone by cartilage, allowing the chest cavity to expand and contract during breathing
13th rib	floating rib – only attached to the spinal column

(especially of the thoracic spine), the diaphragmatic vertebra (can be felt as a dent) and the sacral vertebrae (which fuse within the first two years to form the sacrum).

Ribs should also be easily palpable, so if it is difficult to find them the dog is probably overweight.

The number of ribs corresponds with the number of thoracic vertebrae, with each rib being attached in front of its relevant vertebra. The first pair of ribs is therefore located near the last neck vertebra. The attachment sites of the ribs to the vertebrae form small joints. Each rib consists of a bony and a cartilaginous part.

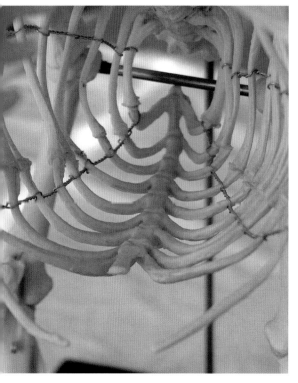

The ribs attach to the breastbone. This is the view from the tail end into the chest cavity. (Photo: Mielke)

The breastbone of a dog comprises eight different bones; it remains flexible for most of the dog's life and only calcifies in old age.

The skull is divided into the facial part and brain (cranial cavity) part. Unlike humans, the dog's facial part is larger than the brain area owing to the more developed jaws and chewing muscles, the exception being extremely short-faced breeds such as pugs or bulldogs.

An adult dog has a total of 42 teeth, 20 of which are situated in the upper jaw and 22 in the lower jaw. The incisors are the small teeth at the front of the mouth; carnassial teeth for tearing food apart are located near the back.

The tooth surface is made of enamel, the hardest substance in the body.

The dog's teeth are adapted to its role as a predator. (Graphic: Maehler)

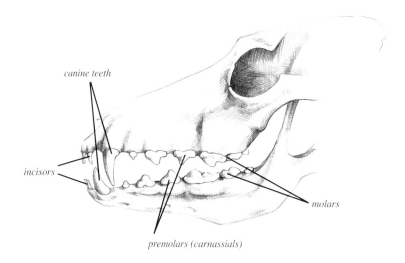

canine teeth

incisors

molars

premolars (carnassials)

Puppies are born toothless and by the age of six weeks all 28 deciduous (temporary) teeth are present. Teething takes place between three and seven months of age depending on breed.

The joints

Joints are divided into true and false joints. Connection of parts of the skeleton through muscles or connective tissues such as ligaments is defined as a false joint. The shoulder blade, for example, is only connected to the trunk by muscles; there are no bony joints.

True joints are also called synovial joints. The terminology of a joint is based on the number of bones involved and the shape of the joint. They all have a gap between the articulating bones, thereby allowing movement. A simple joint consists of just two bones (e.g. the shoulder joint) whereas a complex joint involves three or more bones (e.g. the stifle joint). There are various types of true joint.

The gliding joint has two flat articular surfaces that are connected by strong ligaments. The range of movement is limited. Such a joint is formed between the articular processes of vertebrae.

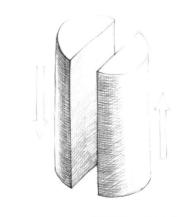

Sketch of a gliding joint.

The ball and socket joint has a rounded articular surface that sits inside a socket formed from another bone. Theoretically this joint can be moved in every direction, but the ligaments and muscles that hold it together limit the movement. The shoulder joint is an example of a ball and socket joint.

Sketch of a ball and socket joint. (Graphics: Maehler)

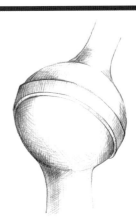

A special form of a ball and socket joint is the hip joint, where the socket reaches far over the head of the ball.

The ellipsoid joint consists of a raised elliptical joint surface and an equivalently shaped socket. The head and neck are connected by such a joint.

Sketch of a hip joint.

Sketch of an ellipsoid joint.

The pivot joint consists of a fixed shaft surrounded by a hollow moving cylinder. The joint between the first and second neck vertebrae is an example of this.

The saddle joint consists of two reciprocal saddle-shaped elements that permit only flexion and extension. The claw joint of the dog's paw is an example.

Sketch of a pivot joint.

Sketch of a saddle joint. (Graphics: Maehler)

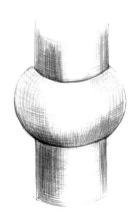

The hinge joint has the shape of a barrel with a surrounding joint surface.

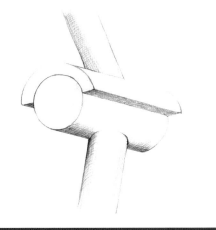

Sketch of a basic hinge joint.

A hinge joint with a screw-like function also sits on a barrel but at an angle to the plane of movement, permitting slight lateral movement as well as flexion and extension. The hock is such a joint.

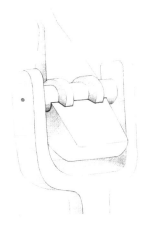

Sketch of the type of hinge joint found in the hock.

This type of joint also only permits flexion and extension. There are different types of hinge joint. Depending on their shape they act like a hinge, screw, sledge or spiral. An example of a hinge joint in the dog is the elbow joint.

Sketch of the type of hinge joint found in the elbow joint.

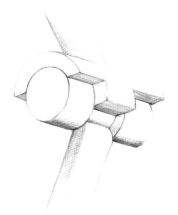

A hinge joint shaped like a sledge has a bone that sits like a bobsleigh on a track. The track – the lower joint surface – is shaped by barrel-like structures, called condyles. An example of such a joint is the patella joint.

Sketch of the type of hinge joint found in the patella joint. (Graphics: Maehler)

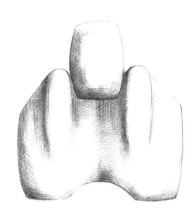

A spiral hinge joint has convex (curved to the outside) articular surfaces. The degree of curvature increases towards the rear of the joint. When the joint is flexed the lateral ligament tightens thus stopping the flexion. An example of such a joint is the stifle joint.

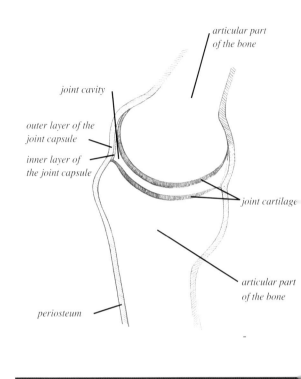

Components of a true joint. (Graphics: Maehler)

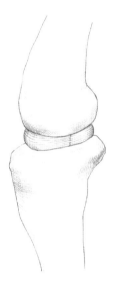

Sketch of a spiral hinge joint, as found in the stifle joint. (Graphic: Maehler)

The components of a synovial joint are the cartilage, the joint cavity and capsule, ligaments, joint fluid and intra-articular structures.

The joint cartilage is formed of hyaline cartilage. It is only a few millimetres thick and has a very smooth surface. It is firmly attached to the bone. The word hyaline stems from the Greek and means 'glassy'.

Hyaline cartilage consists mainly of collagen fibres (see also the section on bones) and has a milky bluish colouration. It is devoid of blood vessels and therefore needs other means of nutrient supply. Cartilage depends on the diffusion of joint fluid (synovia) for its nutrients, which are produced by the inner layers of the joint capsule. Diffusion is the transport of molecules from a higher concentration towards an area of lower concentration until the concentration is equal in both parts. In a joint the pressure exerted on the joint surfaces helps to press the fluid into the cartilage.

Disturbances of cartilage nutrition can result from a lack of exercise as well as from over-exertion. The right amount of exercise is important.

If the cartilage is not supplied sufficiently with nutrients it can break down, which leads ultimately to chronic arthritis. Arthritis is an inflammation of the soft tissue parts of a joint (cartilage and capsule) that can lead to very painful and immobilising new bone growth along the bony parts of the joint. Chronic arthritis is incurable. The symptoms can at best be alleviated and the disease progression slowed, because hyaline cartilage only has a limited ability to recover from injury.

The joint capsule consists of two layers. The outer layer is made of strong connective tissue, that gives the joint stability. The inner layer is well supplied with blood vessels, nerves and lymph vessels. It has innumerable small projections on its inner surface that produce the joint fluid. Synovial fluid is a clear, slightly yellow fluid. It prevents friction of the joint cartilage during movement and, as mentioned before, supplies the cartilage with nutrients.

Joint fluid is viscous. Its quality is defined by the amount of hyaluronic acid dissolved in it. Hyaluronic acid can adjust to the required viscosity of its surroundings; this process is called thixotropy. The principle is similar to using plasticine: once plasticine has been kneaded for a while it becomes softer and can be shaped easily. When left alone for some time it hardens again. This is similar to what happens inside a joint. If strong forces act on the joint (e.g. when standing) the joint fluid thickens, but when the joint is exercised during running the fluid viscosity decreases, thus reducing friction. An injured joint (caused by trauma, inflammation or wear) produces larger amounts of joint fluid. The joint then becomes extended (hydrops). The synovial fluid in such a joint is usually watery, sometimes even opaque and dark and has lost its cushioning qualities.

Joint ligaments can be located inside as well as outside the joint capsule. They connect the articular bones with each other and guide their movement.

Other intra-articular structures include cartilaginous discs such as the menisci in the stifle joint. They are essential because the joint surfaces are incongruent. In other words, the surfaces do not fit together precisely and the joint requires a cushion in order to function properly. The two menisci in the stifle have shock absorbing properties. Meniscus means 'small moon' and refers to the crescent-like shape of the discs.

Selected joints and their functions

• Shoulder joint

The shoulder joint is a ball and socket joint with a theoretical movement in all directions. However, ligaments and muscles limit movement in a dog's shoulder and rotation is only marginally possible. The joint surfaces are formed by the concave end of the shoulder blade and the rounded end of the humerus.

The humerus does not have the typical lateral ligaments. Their function is taken over by elastic tendons that originate from the muscles on either side of the shoulder blade. The tendon of the long head of the biceps muscle originates from a protrusion of the shoulder blade and passes within the joint capsule, which acts as a tendon sheath. This makes the area prone to injuries and inflammation.

Between the tendon of the supraspinatus muscle and the joint capsule lies a bursa, a small soft cushion that alleviates pressure forces.

• Hip joint

The hip joint is a ball and socket joint that allows a wide range of movement. The joint is formed by the convex end of the femur and a concave socket in the pelvis, the acetabulum. Three pelvic bones that are fused with each other form the socket: the ilium, ischium and pubic bone. An intra-articular ligament holds the femur in place. The hip joint is surrounded by large muscles that also offer stability.

Large breeds of dog are prone to developing hip dysplasia. With this condition, the femoral head and socket do not fit correctly together because the bones are poorly shaped. The degree can vary from a slight misalignment to grossly deformed femoral heads and completely flattened sockets. The degree of hip dysplasia often does not correlate with the degree of pain experienced by the dog. Hip dysplasia always results in chronic arthritis, so-called coxarthritis. The instability of the affected hip joint leads to disturbances of gait and movement, which also results in increased wear of the articular cartilage. The joint capsule develops inflammation, which causes thickening, shrinkage and diminished circulation in the

The shoulder joint. (Graphic: Maehler)

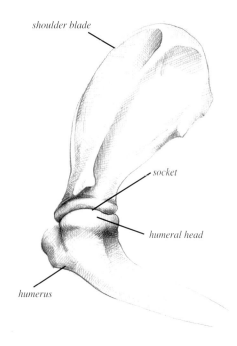

shoulder blade

socket

humeral head

humerus

capsule. At the same time, the bony edges of the joint react with additional bone growth that can give the appearance of the surface of a cauliflower. It is assumed that the body tries to compensate for the reduced stability through additional bone growth.

The hip joint as viewed from the side. (Photo: Mielke)

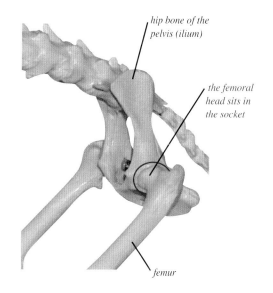

hip bone of the pelvis (ilium)

the femoral head sits in the socket

femur

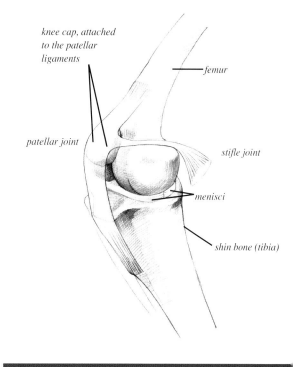

knee cap, attached to the patellar ligaments

femur

patellar joint

stifle joint

menisci

shin bone (tibia)

Sketch of a stifle joint. (Graphic: Maehler)

• Stifle joint

The stifle joint consists of the femorotibial articulation (between the thigh and shin bone) and the femoropatellar articulation (between the thigh and knee cap). Both parts of the joint work in conjunction with each other, never alone. The stifle joint is an incongruent spiral joint. The joint surfaces of the thigh and of the shin bone do not fit exactly, thus requiring two menisci for support. The cruciate ligaments prevent the joint surfaces from sliding too far forwards or from extending too far backwards. A torn anterior cruciate ligament is one of the most common orthopaedic problems in dogs.

• Patellar joint

The patellar joint is a hinge joint shaped like a sledge. The joint is formed by the knee cap and the condyles of the thigh. The knee cap slides up and down over the rounded joint surfaces of the thigh, thus distributing the active forces in a new direction; the pressure on the upper thigh mus-cles is redistributed on to the calf muscles when the knee is bent.

The skeletal muscles

The main function of the skeletal muscles is to produce movement, but they also sta-bilise the joints and enable the dog to stand. Standing is an active process, which explains why dogs tend to lie down when-ever possible. Muscles can also produce heat by shivering, which involves rapid mus-cle contractions. Skeletal muscles form the large body walls and they support organ functions, e.g. through breathing move-ments or abdominal contractions. Such contractions are essential for passing fae-ces and urine, for vomiting and for giving birth.

When looking at muscles under a micro-scope a fine striation can be seen, which

Structure of a muscle cell. (Graphic: Maehler)

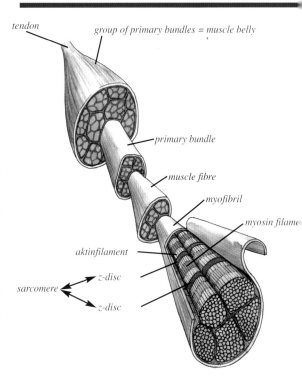

tendon

group of primary bundles = muscle belly

primary bundle

muscle fibre

myofibril

myosin filament

aktinfilament

z-disc

sarcomere

z-disc

*Sketch of a sarcomere.
(Graphic: Maehler)*

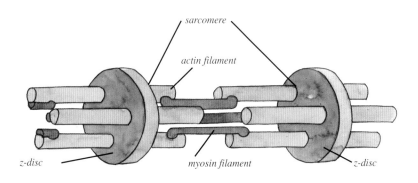

sarcomere

actin filament

z-disc

myosin filament

z-disc

is caused by the various densities of the small fibres, also called filaments. The smallest contractile element of a muscle is the sarcomere. A sarcomere is part of the myofibril. Many hundreds of myofibrils form one muscle fibre or muscle cell; many muscle fibres together are called a primary bundle; a group of primary bundles makes up the muscle belly.

The Latin name for muscle is musculus. Literally translated this means 'little mouse'. When looking at the shape of a muscle belly with its tendon it is indeed reminiscent of a mouse – it has a head, belly and tail.

The main function of skeletal muscles is to enable movement. In order to move the bony parts of the skeleton, muscles need an origin and an attachment and when these move the dog, or part of the dog's body, they contract. Some muscles only move the bones of one joint but others act on several joints at the same time. For example, the large upper thigh muscle (quadriceps) can extend the knee as well as flex the hip.

Skeletal muscle movement is controlled actively by the dog. The brain sends the command to lift a paw because the dog knows that it will be rewarded with a treat for that action. The command is transmitted via nerves stimulating the relevant muscles to contract. Several joints are then moved in order for the dog to lift its paw.

The tendinous origin of the muscle is attached to an unmovable part, usually closer to the body. The tendinous insertion is attached to a movable part, which is usually further away from the body. In some cases this can be reversed, with the moving part being close to the origin of the muscle.

As mentioned above, a muscle has a tendon at both ends. These tendons can have very different shapes. Spindle-shaped muscles end in rope-like tendons, whereas flat, wide muscles end in broad tendon sheets. The tendinous attachment to the bone is very strong. A muscle can never tear a healthy tendon! When a tendon is exerted excessively it is more likely that a small piece of bone will be torn off than the tendon being torn.

Components of a muscle. (Graphic: Maehler)

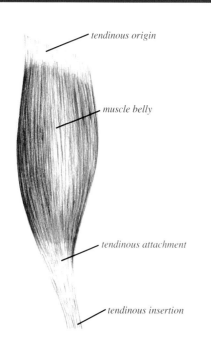

tendinous origin

muscle belly

tendinous attachment

tendinous insertion

However, permanent over-exertion can weaken a tendon and such a damaged tendon can then tear with even normal workload.

Depending on their function muscles can have various shapes – spindle-shaped, rope-like or sheet-like. Spindle-shaped muscles are mainly located on the legs. Along the spine the dog has rope-like muscles, whereas the muscles of the body wall are broad and wide. The largest muscles of a dog are located in the rump and hindquarters.

Muscles can have several heads (one to four). Depending on how the tendon invades the muscle or surrounds it, the muscles are called pennate, bipennate or multipennate. A multi-pennate muscle can develop great strength but contracts very little. Such muscles are best suited for continuous workload, e.g. stabilising joints while standing. A simple pennate (unipennate) muscle shortens more during contraction but its strength is less and it wears out more quickly.

Muscles that move across the flexing side of a joint are called flexors. They reduce the joint angle during contraction. On the other hand, extensors increase the joint angle during contraction thus extending a joint. A muscle that pulls a leg sideways towards the body is called an adductor and one pushing the leg outwards is an abductor.

Muscles can also be divided into synergists and antagonists. Synergists work together during contraction whereas antagonists pull in different directions. When both exert the same force on a joint they have a stabilising effect. All the different functions are important to allow variation of movement. For example: the two-headed biceps muscle of the upper foreleg contracts during flexion of the elbow joint. The brachialis muscle acts as a synergist. The triceps muscle is their antagonist because it causes elbow extension and has to relax during flexion.

Several accessory structures facilitate the smooth movement of muscles across bones and against each other. Muscle fascia consists of a thin layer of connective tissue wrapped around a muscle or muscle group, allowing the muscles to slide against each other without sticking. Fascia also connects the superficial tissues such as skin and skin muscles to deeper structures such as bones and skeletal muscles.

Bursae, tendon sheaths and sesamoid bones protect the muscles and tendons that are located close to bony protrusions. A bursa is a pouch with two outer layers filled with a fluid that is similar to joint fluid. A tendon sheath is a bursa shaped like pipe insulation surrounding a tendon. The tendon lies inside like a knife in a sheath.

Almost all muscles exist in matched pairs located on the left and right sides of the body.

If you notice that a muscle on one side of your dog is better developed than the one on the other side you should ask a vet to check your dog. A muscle may not be used correctly owing to an underlying problem and will show signs of wasting (atrophy). Muscle atrophy develops quite quickly and can be visible after as little as a few days in cases of complete disuse.

How does a muscle contract?

First of all the muscle has to be given the command to contract, which is transmitted by nerve cells. A nerve cell controlling a particular muscle is called a motor neuron and it works in unison with the relevant muscle. The signal is transmitted via a synapse. The exact biochemical process of nerve transmission goes far beyond the boundaries of this book. To keep it simple we accept that the stimulus reaches the muscle.

I want to return to the sarcomere, the smallest contractile unit of the muscle. The sarcomere consists of actin and myosin filaments located between the z-discs. Each myosin filament has two heads that slide across a kind of neck. During contraction the myosin heads attach to the actin filaments. They then

bend (up to 45 degrees) thus shortening the sarcomere at both ends.

This process is called the 'sliding filament theory'. Following this action the attachments are disconnected, the heads are extended and the whole process starts again. It might help to compare this to a tug-of-war. Tug-of-war is a process of pulling, relaxing and pulling repeatedly until the goal is reached.

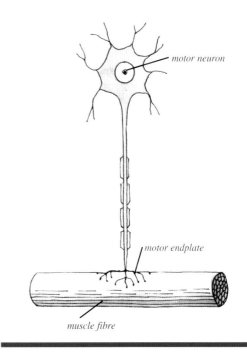

motor neuron

motor endplate

muscle fibre

A nerve cell and muscle fibre connect at the synapse. (Graphic: Maehler)

One sliding cycle shortens a muscle about one per cent. This means that for many movements repeated cycles are necessary: attaching the heads, bending and sliding, disconnecting, extending,

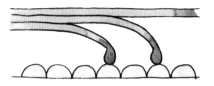

myosin heads attach to the actin

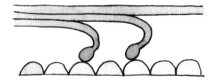

the sliding movement causes contraction

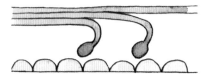

disconnecting the attachment

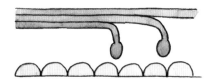

extending the heads

Sketch of the sliding filament theory.
(Graphics: Maehler)

attaching at another point of the actin filament, and so on. Energy is needed for attaching as well as disconnecting. This explains rigor mortis: there is no energy left in the body to disconnect the myosin heads.

A healthy muscle has a basic tension, also called muscle tone. A muscle is never completely slack! It is always on call ready to trigger a movement. If the nerve supplying a particular muscle is permanently damaged or cut, the muscle can no longer contract and relaxes completely (like a piece of steak before frying). A muscle can also have in-creased tension for many reasons (e.g. illness or exertion).

> You can detect tense muscles in your dog because it finds them painful to your touch. When gently exploring the sore area you can feel a hardened muscle. It feels a bit like a guitar string.

Muscles important for movement

Let's look at the front leg first. The movement of the front leg involves lifting it off the ground, backwards movement, extending forward, and placing it back on the

Lifting the front leg off the ground with a slight backwards movement. (Photo: Mielke)

ground. Lifting and backwards motion involves the following joint movements: flexion of the shoulder, extension of the elbow and flexion of the lower leg.

The most important muscles involved in lifting the foreleg

Joint movement	Muscles responsible for the joint movement
Flexion of the shoulder	latissimus dorsi
	deltoid
	teres minor (small rounded muscle)
	teres major (large rounded muscle)
	triceps
	subscapularis
	rhomboid
	infraspinatus
Elbow extension	triceps
	anconeus
	tensor fasciae antebrachii
Flexion of the lower leg	flexor carpi ulnaris
	flexor carpi radialis
	extensor carpi ulnaris
	flexor digitorum longus muscles (long toe flexors)

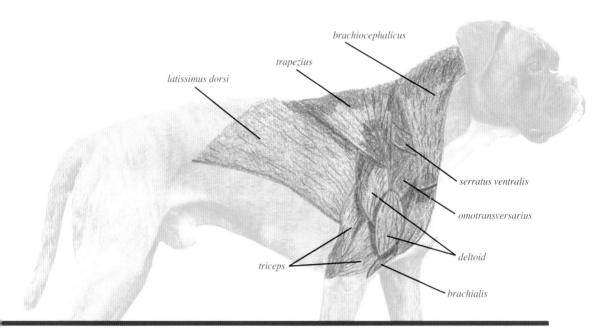

brachiocephalicus

trapezius

latissimus dorsi

serratus ventralis

omotransversarius

deltoid

triceps

brachialis

Muscles of neck, trunk and front leg. (Photo and graphic: Mielke)

Outer (lateral) muscles of the front leg.

Inner (medial) muscles of the front leg. (Graphics: Maehler)

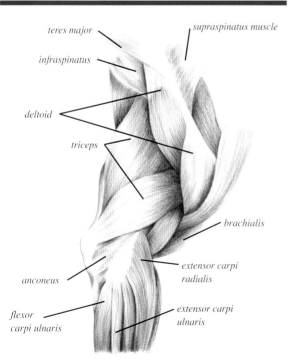

teres major

infraspinatus

supraspinatus muscle

deltoid

triceps

brachialis

anconeus

extensor carpi radialis

flexor carpi ulnaris

extensor carpi ulnaris

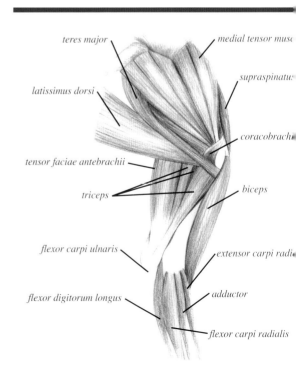

teres major

medial tensor musc

latissimus dorsi

supraspinatu

coracobrach

tensor faciae antebrachii

triceps

biceps

flexor carpi ulnaris

extensor carpi radi

flexor digitorum longus

adductor

flexor carpi radialis

Moving the front leg forwards and placing the foot back on the ground.
(Photo: Mielke)

For the forward movement of the front leg and placing of the foot back on the ground the following joint movements are needed: shoulder extension, elbow flexion and extension of the lower leg.

The following muscles of the shoulder area are involved in these movements: the trapezius muscle lifts the front leg and the brachiocephalicus muscle moves the front leg forward. In other words, the

The most important muscles involved in placing the forefoot on the ground

Joint movement	Muscles responsible for the joint
Shoulder extension	biceps
	supraspinatus
	subscapularis
Elbow flexion	biceps
	brachialis
Lower leg extension	flexor carpi radialis
	extensor digitorum longus muscles (long toe extensors)

brachiocephalicus is the most important muscle in the forward motion of the front leg.

As mentioned earlier, the shoulder blade and trunk are not connected by a true joint but by a false joint made of muscles and ligaments. The muscular attachment allows better absorption of vertical forces. The serratus ventralis muscle acts as a 'trampoline' and carries the trunk. This muscle contains a large amount of tendinous fibres and is therefore able to carry significant weight with a minimum of effort.

Other muscles are also part of the attachment of the front leg to the trunk: the superficial and deep pec-toral muscles, latissimus dorsi, brachiocephali-

Muscles of the trunk and chest.
(Photo and graphic: Mielke)

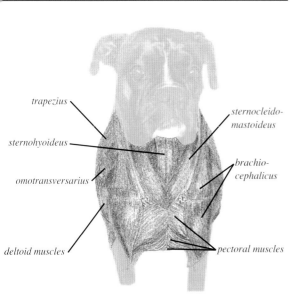

cus, trapezius and the rhomboid muscle. The front leg can also be moved sideways. The deltoid and infraspinatus muscles pull the leg outwards; the pectoral and coracobrachialis muscles pull it towards the body.

You may have noticed a slight up and down neck movement when your dog gallops. It is easier to see in horses because of their larger size but dogs do the same. These movements help to balance the animal and are caused by the splenius capitis, a strong, broad muscle in the upper neck area.

The longissimus dorsi muscle extends and stabilises the spine.

When looking at the hind leg, its movements also involve lifting off the ground, backwards movement followed by forward motion, and placing back on the ground. The placement on the ground, however, is a more powerful and bouncing action than that of the front leg. The following joint movements are involved: extension of the hip, flexion of the stifle and flexion of the hock.

Placing the foot and pushing the hind leg forward involves the following joint actions: flexion of the hip, extension of the stifle joint and extension of the hock.

The dog can also push the hind leg outwards from the hip joint. The gluteus muscles, pirifiormis, biceps femoris and the abductor muscles enable this movement.

Pulling the leg towards the body involves the gracilis muscle, pectineus, adductor muscles, sartorius and semimembranosus.

Lifting and backwards movement of the hind leg. (Photo: Mielke)

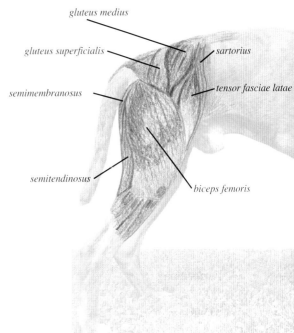

gluteus medius

gluteus superficialis

sartorius

semimembranosus

tensor fasciae latae

semitendinosus

biceps femoris

Muscles of the rump and hind leg.
(Photo and graphic: Mielke)

The most important muscles involved in lifting the hind leg

Joint movement	Muscles responsible for the joint
Hip extension	gluteus superficialis
	gluteus medius
	gluteus profundus
	piriformis
	biceps femoris
	semimembranosus
Stifle flexion	biceps femoris
	semitendinosus
	gastrocnemius
	popliteus
Hock flexion	tibialis cranialis
	fibularis longus
	fibularis brevis
	extensor digitorum longus

Placing the leg on the ground and pushing forwards. (Photo: Mielke)

The most important muscles involved in placement of the hind foot on the ground

Joint movement	Muscles responsible for the joint
Hip flexion	iliopsoas
	quadriceps
	sartorius
	pectineus
	stensor fasciae latae
Stifle extension	quadriceps
	sartorius
	tensor fasciae latae
	biceps femoris
	semimembranosus
	semitendinosus
	gracilis
Hock extension	biceps femoris
	gastrocnemius
	semitendinosus
	extensor digitorum longus

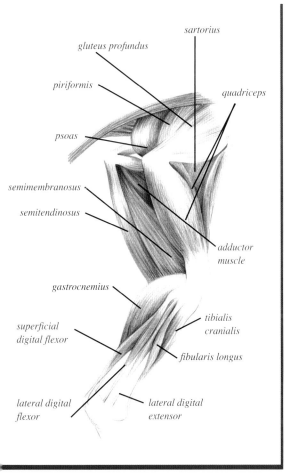

Outer (lateral) muscles of the hind leg.

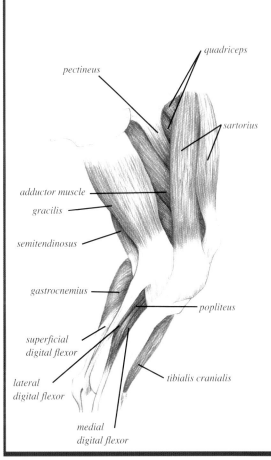

Inner (medial) muscles of the hind leg.
(Graphics: Maehler)

The largest and/or most important muscles that enable movement are summarised in the diagram above.

Each muscle has at least one origin and attachment, at least one function and one relevant nerve. The nerve controls and coordinates movement by stimulating the muscle (see the chapter on the nervous system).

The brachiocephalicus muscle is the main player when moving the front leg forwards. This muscle consists of three parts. It attaches to the front leg as well as the upper neck and moves through the rudimentary clavicle, a tendinous area at the height of the shoulder joint. It is innervated by the axillary nerve. As well as extending the front leg it can also flex the head and neck downwards when the front leg is in a stationary position, as well as bending the neck and head sideways.

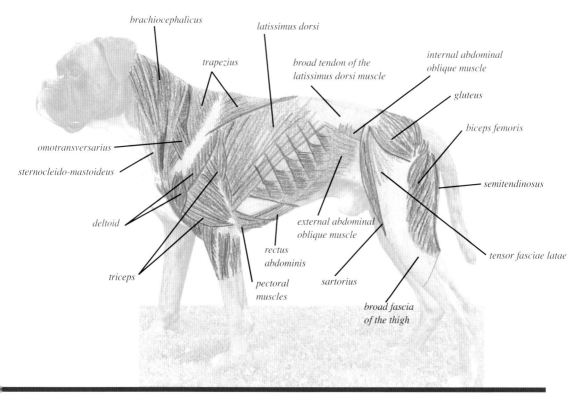

Overview of the superficial muscles – for quick reference.
(Photo and graphic: Mielke)

The latissimus dorsi is the strongest muscle that pulls the front leg backwards. It is also one of the largest muscles in the body. It originates from the spine and attaches to the upper front leg. It is innervated by the thoracodorsal nerve. As well as flexing the shoulder joint it also pulls the trunk forward when the front leg is on the ground. The triceps brachii muscle actually has four heads in the dog. It makes up the triangular area of the upper arm. The triceps muscle originates from the shoulder blade and the upper front leg and attaches to the tip of the elbow. It extends the elbow joint and flexes the shoulder joint. The radial nerve supplies this muscle.

One of the strongest muscles in the body is the medial gluteus muscle. It originates from the side of the ilium as well as the last sacral and first tail vertebrae and attaches to the femur. It is a strong hip extensor and also enables outward movement from the hip joint. The cranial guteal nerve is its relevant nerve.

The two-headed biceps femoris muscle is another one of the strongest muscles of the dog. It lies under the skin and reaches from the seat bone to the middle of the lower thigh. It extends the hip joint, stifle and hock. During running it can also flex the stifle and it can push the hind leg outwards. It is innervated by the caudal gluteal and tibial nerves. The pectineus muscle is a small spindle-shaped but very strong muscle that originates from the pubic bones and attaches to the middle of the femur. It flexes the hip and pulls the leg towards the body, holding the head of the femur in the hip socket. Dogs with hip deformities (hip dysplasia) overly exert this muscle, causing tension, pain and increased wear of the joint cartilage. The pectineus muscle is supplied by the obturator nerve.

The large and powerful four-headed quadriceps muscle is located at the front of the thigh. It originates at the ilium as well as three locations on the femur. All four heads join, and the muscle ends in the straight patellar ligament, which attaches at the tibia. The knee cap is inserted in the patellar ligament. The quadriceps muscle is the strongest extensor of the stifle joint. It also stabilises the stifle, and thus the whole leg, and plays a role in the flexion of the hip joint. The femoral nerve innervates this muscle.

The gastrocnemius muscle is a strong, spindle-shaped muscle. It originates with two parts at the lower end of the femur and ends at the Achilles tendon, which is easily visible in a dog above the hock. It is a powerful extensor of the hock but is also involved in flexion of the stifle. It is supplied by the tibial nerve.

The Achilles tendon is particular easy to see in short-haired dogs. (photo: Mielke)

Achilles tendon

How does the dog move?

Unlike people, the dog walks on the pads of its toes, i. e. is it therefore digitigrade, as is the cat. This facilitates speed and ease of movement. The rear part of the body develops the momentum for forward movement. The joints of the hind leg extend and the legs are pushed firmly into the ground, which moves the trunk towards the front legs. The spine is firmly attached to the pelvis through the sacrum in order to transmit the momentum from the hind legs to the front. Extension of the hock, stifle and hip leads to the hind legs pushing off from the ground. The hip, hock and stifle are subsequently flexed again, thus bringing the hind legs forward. Hock, stifle and eventually hip are then extended to allow the leg to be placed back on the ground.

During the support phase (when a hind leg is on the ground) the long back muscle (longissimus dorsi) stabilises the spine. It carries the trunk, thus allowing the shoulder to move forward freely. The front leg supports the trunk without playing an active role in forward movement. The shoulder and elbow joint extend and push the body upwards; the lower front leg is then extended massively. Flexion of all the joints allows the front leg to lift off the ground. The brachiocephalicus muscle (leg-head muscle), with help from the trapezius and omotransversarius muscles, pulls the suspended front leg,

including the shoulder blade, forwards. The joints are then extended and the leg is placed back on the ground.

A dog has the following gaits: walk, trot, pace and gallop. A jump is a special form of gallop. During the slowest gait, the walk, the coordination between front and hind legs can be observed quite easily. A dog walks like two people behind each other. These two people have the same rhythm but when one moves their right leg, the other one moves their left leg, with the person at the rear leading by half a stride. The whole cycle therefore is slightly out of sync.

The trot is faster and can be sustained for long periods. A freely running dog usually trots. Using the same analogy as above, the time difference between the two moving people is less in trot. Some breeds of dog move the diagonal pairs of legs at exactly the same time. Like horses, they have a short phase of suspension in between cycles, when the legs are completely off the ground.

During the pace the dog moves the left hind and front legs forward at the same time, followed by both right legs. This time the two people march in lock-step. To pace is considered a fault in some breeds, whereas for other breeds it is a standard gait. For example, the pace is undesirable in a Boxer but is a standard gait for a Springer Spaniel and it is tolerated in an Old English Sheepdog. If a dog suddenly adopts a pacing gait it can be a sign of a locomotor problem.

It is easy to see here that the momentum stems from the hind legs.
(Photo: Lehari)

The front legs support the trunk, as you can see in this photo. (Photo: Lehari)

The gallop is the fastest gait. The top speed attainable depends on the breed of dog, with Greyhounds probably being the fastest, achieving speeds of up to 60 kilometres per hour. The gallop is a jumping motion with the dog jumping from the hind legs to the front legs. The hind legs thrust the trunk forward with great momentum and the front legs catch it. If the spine were completely rigid the gallop would not be possible, therefore the flexibility of the spine plays an essential role. Jumping across an obstacle is a form of gallop and usually develops from it, although some dogs can jump from a standing start if the obstacle is not too big.

Dogs do not like to walk backwards. They always prefer to turn around and walk forwards again but they can be taught to go backwards. The reason for a dog's reluctance to move backwards is that its skeleton is not well suited to this action. When reversing the momentum has to come from the front legs, but the muscle strength is insufficient and the lumbar and sacrum area is too rigid. Everything that is perfect for forward movement is a disadvantage for reversing. If a dog has to step backwards it holds its head quite high in order to push its centre of gravity backwards.

As in humans, a dog knows the position of its legs even with its eyes closed. Active and passive joint movements can be felt without having to be seen. Sensory receptors in muscles, tendons and joints enable this. These receptors are called proprioceptors. They are specialised cells that gather information about stimuli acting on the body, and they determine how to deal with them.

Information about the position of the legs and about muscle tone is collected by muscle spindles, Golgi organs in tendons and the Ruffini corpuscles. Muscle spindles are specialised sensory muscle cells that regulate muscle length and therefore measure the exact force of the muscle. They also prevent overstretching. Sensory cells in tendons, named Golgi sensors after their discoverer, measure the tension of the tendons and muscles. The Ruffini corpuscles, also named after their discoverer, fulfil the same function but are located in ligaments and joint capsules. They also transmit information about the angles of the joints.

Are coordinated movements inherited?

No – they are not! This becomes obvious when observing puppies. Their clumsy and uncoordinated movements are what makes them so endearing and entertaining to watch. The motor abilities of a newborn pup are limited and most movements have to be learned. Although the nervous system has developed all the required wiring at birth, the functions of the muscles have to be activated and practised in order to establish organised smooth movements.

The essential element for learning a new movement is the correct timing of muscle contractions and subsequent joint actions, i.e. good coordination. Motor ability can be assessed by measuring the consumption of oxygen and energy associated with a specific muscle movement. A dog that swims for the first time will use up significantly more energy than a dog that is an experienced

Dogs, like these 20-day-old Boxer pups, have to learn most movements by practising. (Photo: Mielke)

With the right training jumps like this present no problem. (Photo: Lehari)

swimmer. A dog that has already learned to swim will have more coordinated and efficient movements. They will not be struggling to stay above the water but will swim with focused and strong movements.

The basis for all skills is the correct training. The phrase 'No one is born an expert' can also be applied to dogs while they are adjusting to the new activity and until they have perfected it. After that the new skills need to be maintained within the genetic possibilities of the individual. Even the most comprehensive training will not teach a dog to fly!

> Dogs also suffer from sore muscles after unusual or extremely hard exercise, but they cannot easily tell us that they ache.

Jumping has to be practised, too. A dog should not be introduced to jumping before it is physically mature (around one year of age), and you should start with small fences, increasing the height gradually. A fully trained dog, e.g. for agility training, does not have to jump the maximum height at every training session. The aim of training is often to improve general technique, and smaller fences are sufficient for this. Your dog will still be able to clear a bigger fence on competition day!

Where does the energy for muscular activity come from?

The energy for muscular work stems from the dog's digested food. A balanced species-specific diet is the basis for a sufficient energy supply for powerful, sustainable movements. But how does food reach the muscle? The digestive and circulatory systems extract energy from food and distribute it throughout the body, where it is stored. The storage capacity, and therefore energy availability, differs from dog to dog. Different energy carriers can be mobilised depending on the intensity of the exercise, with fluid transitions between them.

Muscle can only use adenosine triphosphate (ATP) as fuel for contractions. However, only small amounts of ATP are stored in the muscle. Therefore ATP has to be produced continuously. Three different methods of ATP production are listed in the table below, and their activation depends on the type of exercise.

Metabolic processes of ATP production

Splitting of creatine phosphate	this is fast, and is used for short intense exercise; it occurs directly in the muscle stores; creatine is produced in the liver and kidney and binds to phosphate in the muscle to be available when needed
Anaerobic glycolysis	this takes slightly longer, and is used for longer lasting intense exercise; it involves transformation of carbohydrates when limited amounts of oxygen are available
Aerobic glycolysis	this is slow, and is used for continuous exercise; it involves transformation of carbohydrates and fatty acids with the help of oxygen

The disadvantage of anaerobic glycolysis is the subsequent build-up of lactic acid in the muscle cells. The lactic acid has to be metabolised further. It increases the cell's acidity (lowers the pH), thus blocking the chemical processes necessary for muscle contraction. The muscle tires as ATP production decreases, until it cannot work any longer. New energy carriers are produced while the muscle is resting.

Consistent training of muscles will increase their energy storage capacity. For example, exercising your dog during a bicycle ride three times a week for 30 to 45 minutes may be ideal. It is of course essential that your dog is healthy and fully grown. You should use a chest harness when leading the dog next to the bicycle and exercise it on soft, elastic ground.

Other muscles

Apart from skeletal muscles a dog has many other muscles. The most important muscle – the heart – is described in a different chapter.

Respiratory muscles

Respiratory muscles can be divided into muscles for inhalation and those for exhalation. Dogs have very strong respiratory muscles, the largest of which is the diaphragm (see graphic in the chapter on the internal organs). During inhalation the diaphragm contracts and extends into the abdominal cavity, thus enlarging the chest cavity. The reduced pressure in the chest expands the lungs and sucks in air; the dog inhales.

Other important respiratory muscles are located between the ribs, along the sternum and the spine. These muscles lift and lower the ribs. The intercostal muscles (between the ribs) are relatively tendinous. Their structure and crisscross arrangement stabilise the rib cage and thus the whole body.

Abdominal muscles

The abdominal muscles consist of a broad muscular sheet composed of four muscles. They are the inner and outer oblique, the straight and the transverse abdominal muscles. These muscles form the abdominal wall and enable a dog to strain, e.g. when passing urine or faeces, as well as during labour. The muscles are

The abdomen of the dog. (Photo: Mielke)

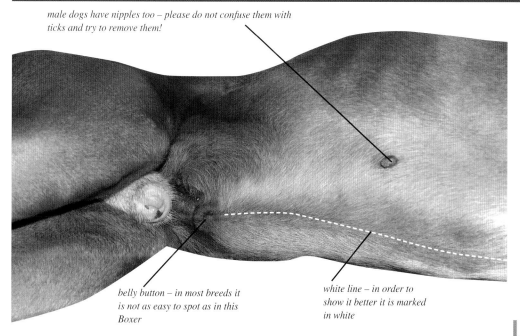

male dogs have nipples too – please do not confuse them with ticks and try to remove them!

belly button – in most breeds it is not as easy to spot as in this Boxer

white line – in order to show it better it is marked in white

also involved in exhalation and in bending of the spine. A tendon-like band of connective tissue, the so-called white line, runs down the middle of the abdomen from the chest bone to the pubic bone. This line includes the belly button, which dogs have as well as humans!

Muscles of the head

The muscles of the head can be divided into facial muscles (flews or lips, eyelids, ear and nose muscles) and chewing muscles (muscles of mastication). The facial muscles are firmly connected to the skin (especially in the area of the flews, nose, ears and eyes) and are also referred to as the mimic muscles. The muscles of mastication are strong and very well developed. In particular, the outer masseter and the temporal muscles contain strong tendinous tissue. They lift the lower jaw and press it against the upper jaw to enable chewing, and also grabbing and holding on to objects with the muzzle, e.g. when retrieving items.

Muscles of mastication. (Photo and graphic: Mielke)

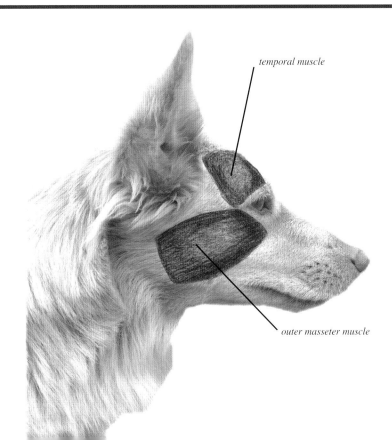

temporal muscle

outer masseter muscle

Skin muscles

The skin muscles are located under the skin. They are surrounded by fascia, i.e. a layer of connective tissue, which is firmly connected to the skin. Most skin muscles do not have an origin and attachment but are spread over large parts of the body. They allow the dog to move individual areas of skin. For dogs the skin muscles are a part of their communication system, because they can contract or fold their skin when required.

Muscles of the tail

There are two groups of tail muscles: the spine-tail muscles and the pelvis-tail muscles. However, they can also be distinguished by their function: to lift or lower the tail or to move it sideways. The tail muscles allow the tail to be carried or moved in very well-defined ways, which again is an important means of communication for the dog. Even just the tip of the tail can be moved. The tail supports the expression of tension, excitement, fear, aggression or joy. The tail also acts as a balancing tool, which is essential when jumping or walking across tree trunks or narrow paths as required for agility training. This follows the same principle as the pole that acrobats use for tightrope walking. Such a balancing pole makes subtle adjustments of position easier and faster. A Boxer breeder even told me that since the ban on tail docking her puppies have been learning to walk sooner.

The Animal Welfare Act 2007 bans tail docking and ear cropping in dogs. After having read the last section I hope you welcome this ban as much as I do, especially if you own a dog of one of the formerly affected breeds.

Internal organs

The internal organs lie well protected within the chest cavity and abdomen. The heart, lungs and the thymus of young dogs are located in the chest area. The thymus is essential for the development of the immune system and is only present in juvenile dogs because it disappears with puberty. The

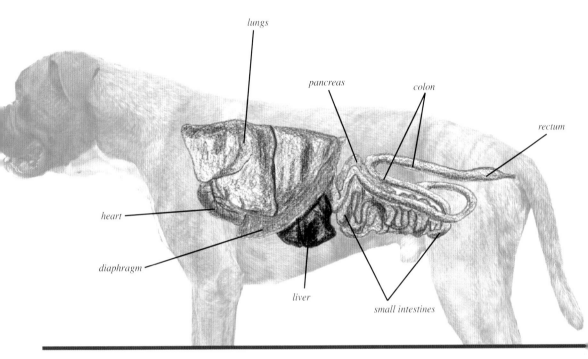

lungs

pancreas

colon

rectum

heart

diaphragm

liver

small intestines

Overview of the internal organs. (photo and graphic: Mielke)

aforementioned organs are enclosed by the chest vertebrae, ribs, sternum and towards the abdomen by the diaphragm (which is also an important respiratory muscle).

In the abdomen lie the intestines, stomach, pancreas, liver, gall bladder, bladder, kidneys, sexual organs and spleen.

The heart is explained in more detail in a separate chapter.

Respiration

In addition to the lungs, larynx, trachea and bronchi, respiratory muscles are needed for breathing (see also the section 'Other muscles'). The largest and most important respiratory muscle is the diaphragm. Respiration is vital because oxygen is the energy carrier essential for the metabolism of every cell. The waste product of the metabolic cycle is carbon dioxide, which needs to be eliminated from the body. All this is accomplished through inhalation and exhalation. Inhalation involves lifting of the ribs, widening of the chest cavity and lowering of the diaphragm, thereby producing low pressure in the lungs and enabling the air to stream in. Exhalation is exactly the opposite; the diaphragm relaxes and the chest cavity decreases in volume, thus increasing the pressure in the lungs. Inhalation is an active process during which the muscles contract, and therefore is more demanding for the dog. Exhalation is usually a passive relaxation of the muscles and only during extreme exercise are muscles involved in this process.

> A resting dog usually breathes between ten and thirty times per minute. Small dogs have a higher respiratory rate than large dogs.

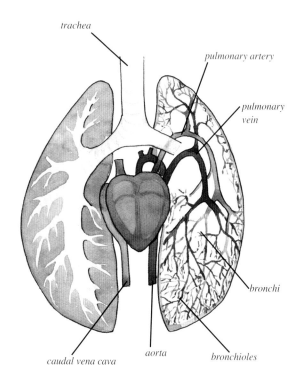

Structure of the lungs. (graphic: Mielke)

Position of trachea and larynx.
(Photo and graphic: Mielke)

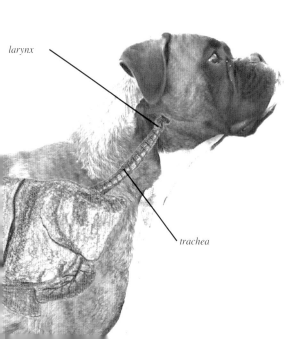

The windpipe (trachea) and oesophagus meet in the larynx. In order to prevent food particles from entering the trachea, the larynx has a lid (epiglottis) that covers the trachea during swallowing. As in humans, the vocal cords of dogs are located within the larynx. A bark is created with the help of the vocal cords and the glottis.

The trachea and larynx are located in the lower side of the neck in a relatively unprotected position.

A chest harness is better for the dog than a neck collar or a chain because of the effect of the latter on the larynx and neck muscles. If the dog wears a neck collar, or even worse a narrow chain, the larynx and upper airways are compressed when the dog pulls.

You can simulate that easily yourself. Press gently with your hand against your larynx; if you increase the pressure this will become uncomfortable. If you press even harder you will most likely begin to cough. Should you press even more the larynx will be squashed. It is the same for the dog.

Dogs that pull on the lead hard can often be heard before they can be seen. The dog tries to tighten the neck muscles in order to alleviate the uncomfortable pressure, causing, as in humans, tension in the upper spine that lead to symptoms such as headaches, dizziness and pain. Unlike humans, dogs are unable to tell us that they are not feeling well, and there is a risk that we overlook the subtle signals they send out.

A chest harness has to be fitted properly in order to prevent rubbing and shoulder problems.

The trachea consists of cartilaginous rings lined with a thin layer of muscle. It moistens and warms the air during inhalation. The trachea splits into two main bronchi and from there into a network of delicate branches (bronchioles) that become smaller and smaller. The smallest units of the lungs are the alveoli, tiny vesicles where oxygen exchange takes place.

The two main lobes of the lungs consist of smaller lobes and segments and are asymmetrical in dogs. The right lung has four lobes and is larger than the left lung, which only has two lobes. The lungs are covered by the pleura, a fine layer that lines the inside of the chest cavity.

Digestion and excretion

The abdomen is separated from the chest cavity by the diaphragm. The abdomen is enclosed by the spine and spinal muscles, the lower ribs and the abdominal muscles. It is lined by the peritoneum, which extends to the organs as well, thus preventing them from sticking to each other.

The oesophagus is built like a muscular tube. It originates from the mouth and ends in the stomach. The latter can be described as a muscular, bag-like extension of the oesophagus. A muscular ring prevents the acidic stomach contents from re-entering the oesophagus. Dogs cannot vomit will-ingly because this function is controlled by the parasympathetic nervous system.

The stomach is lined with a layer of glands. It stores and helps to digest food. Dogs, like humans, have only one stom-ach; they are monogastric. The food is prepared in the stomach and from there it proceeds to the intestines where it is fur-ther broken down, dried out and shaped. The stomach acid of dogs is about ten times stronger than that of humans and a lot more aggressive. It breaks down the food and disinfects it. Dogs have a rela-tively insensitive stomach with regard to bacterial infections as long as the acid content is normal. Including meat in their diet regulates the acidity.

The stomach of a dog lies fairly loosely within the abdomen. The stomach can rotate around its axis, especially in large breeds of dog, which results in the closing of the entrance and exit. This causes a dramatic build-up of gas and distension of the stomach. A rotated stomach is a severe emer-gency that requires immediate surgical correction in order to save the dog's life!

Stomach torsion can be pre-vented by feeding the dog its ration divided into smaller por-tions and letting the dog rest for at least an hour after eating.

	small intestine		large intestine
Duodenum	Here enzymes from the pancreas, gall bladder and liver are added to the food.	Caecum	The dog has no appendix, and therefore cannot suffer from appendicitis.
Jejunum	This is the longest section of the small intestine and is arranged in loops. It is usually completely empty after death.	Colon	Here the fibrous materials that are difficult to digest are broken down.
Ileum	This connects the small and large intestines; its wall contains lymphoid tissue.	Rectum	This is short in dogs. Here water is retained and the faeces are formed.

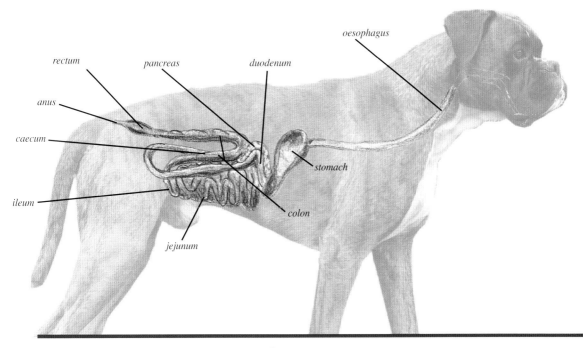

The gastrointestinal tract. (Photo and graphic: Mielke)

The intestines are about five times the length of the dog's body and they end at the anus, which has an inner and an outer sphincter muscle. Near the sphincter muscles lie the anal glands, whose products are important for marking the dog's territory. The anal glands are located in the hairless areas around the anus. They have two pouches that open into the rectum – if you imagine a clock – at four and eight o'clock. Illness or malnutrition can lead to abnormally soft faeces, which prevents the glands from being emptied and causes painful inflammation. A vet then has to treat them.

The liver is involved in detoxifying the body, breaking down hormones and storing vitamins and blood.

The gall bladder is attached to the liver. It produces the bile acids, that bind dietary fats to water thus making them accessible for the body's metabolism.

The pancreas produces enzymes that are needed for the digestion of fat, proteins and carbohydrates. It also produces insulin for the regulation of blood sugar.

A dog has two kidneys, which are located either side of the spine at the level of the first lumbar vertebra. This area should always be kept warm in order not to injure the kidneys. The kidneys filter out all substances from the blood that the body cannot use or that are potentially toxic. These substances are eliminated through the urine.

The urinary system includes the ureter, bladder and urethra. The ureter transports waste products from the kidneys to the bladder, where they are stored and then eliminated via the urethra. The urethra of a bitch is short and wide; that of the male dog is long and thin.

Certain types of nutrient can facilitate the formation of mineral stones in the kidneys or the bladder. Kidney stones are rare, but bladder stones are not. Bladder stones damage the inner lining of the bladder, making passing urine difficult, more frequent and painful. Sometimes the urine contains traces of blood. Owing to the anatomical differences between male and female dogs, blockages of the urethra occur almost exclusively in the male.

Reproduction

The sexual organs of the male dog include the testicles, each with an epididymis, ducts, the prostate gland, urethra and penis. The bitch has ovaries, Fallopian tubes and the womb (uterus).

In the male, the testicles produce semen at sexual maturity, which can begin as early as four months or as late as one and a half years, depending on the breed and the individual animal. The semen is stored in the epididymis and is transported along the seminal ducts to the prostate gland when needed. Inside the penis lies the penis bone (os penis). The bone is surrounded by spongy tissue that swells during mating, thus connecting the dog with the bitch ('the tie') for up to half an hour after ejaculation. The penis bone is not counted as part of the skeleton; it belongs to the organs.

The ovaries of the bitch produce eggs about twice a year, when she is in season. The eggs proceed along the Fallopian tubes to the uterus. The canine uterus consists of two horns and one cervix. After successful mating the embryos lie adjacent to one another in the uterine horns. The cervix only opens during birth or to allow the semen to enter during mating. Sometimes bacteria enter the uterus, causing serious infections (pyometra), especially in older bitches. These infections require immediate treatment by a vet.

Male dogs are always ready for mating, but bitches usually only mate during the tenth to twelfth day of their season. The bitch then does not fight off the dog but accepts him willingly and moves her tail out of his way. Around this time the bleeding from the uterus, which started at the onset of the bitch's season, eases or stops altogether. The eggs also travel from the ovaries into the Fallopian tubes at this time. If the bitch is not mated the uterus returns to its normal state within a few weeks.

Some bitches develop signs of a false pregnancy around this time. They carry

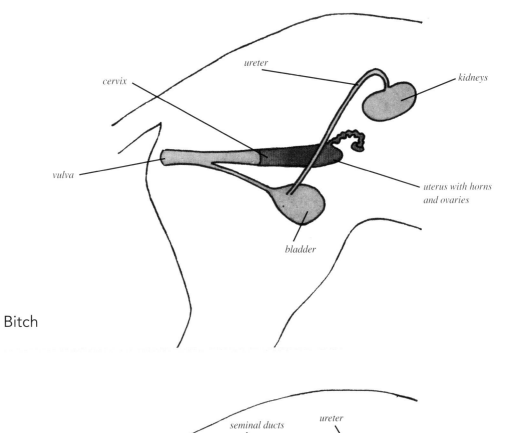

Bitch

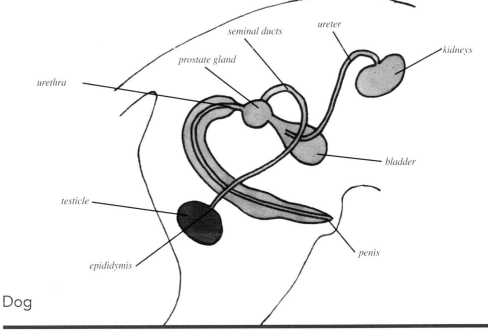

Dog

Overview of the urinary and reproduction organs. Castration of a male dog is a relatively simple procedure: the organs that are removed (the testicles) lie outside the abdominal cavity and are shown in red. In bitches, neutering is a far more complex procedure that involves abdominal surgery. (Graphics: Maehler)

toys, shoes or similar objects into their beds and 'guard' them; some bitches even produce milk. If the symptoms of a false pregnancy are very pronounced you should consult your vet.

If the bitch was mated successfully it takes around 63 days before the birth of the puppies. Nutrients are supplied to the embryos from the yolk sack during the first 28 days of pregnancy. The embryos are particularly exposed to medications or other substances that the bitch may be given.

During the last stage of pregnancy the embryos are fed from the placenta. The foetus develops its head, eyes and extremities first. At around day 33 all vital organs are present. The embryos are still so small that the bitch does not change her body shape.

Another week into the pregnancy the embryos begin to grow significantly and the bitch now begins to fill out. Although the foetuses are completely formed after day 50 they are not yet able to breathe independently. It is at this stage that the mother starts to produce milk.

After 60 days the pups are able to survive and they are usually born on day 63. They are born blind and deaf and detect their mother and her teats using heat sensors. Their eyes open at about 12 to 15 days of age, and after day 17 vision and hearing are fully developed.

Full mental and physical maturity is reached at different ages, depending upon the breed. Generally, smaller breeds mature physically earlier than large breeds. Mental maturity is reached usually by three years of age.

However, the phase of imprinting and socialisation takes place between the third and twentieth week after birth and is the most important period of postnatal development in all dogs. Experiences, or lack of them, during this time will influence the dog's adult life permanently.

Mistakes and undisciplined behaviours are difficult to correct at a later stage. The seventh week is the most crucial; therefore the breeder has a great responsibility while raising the puppies.

Heart and circulation

The heart

The heart is a strong muscle that, like the skeletal muscles, is striped. However, unlike skeletal muscle, its activity is involuntary. The heart muscle is a permanently working pump.

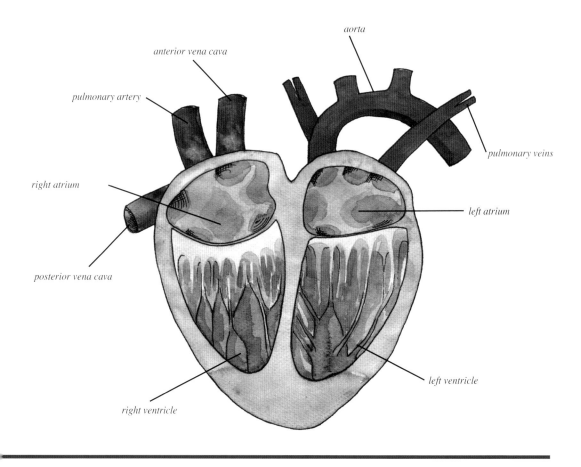

Structure of the heart. (Graphic: Maehler)

The size and weight of a dog's heart depend on several factors: the breed of the dog, its sex, age and constitution. A canine heart weighs between 30 and 500 grams. The heart is located within the chest cavity between the third and seventh ribs. It is not positioned as far to the left as in the human body but only slightly off centre.

The heart is surrounded by a double-walled sac called the pericardium. The pericardium consists of connective tissue and gives the heart its stable position. It also protects the heart (e.g. from lung infections) and minimises enlargement of the heart.

The heart is divided into four sections: right atrium and right ventricle, left atrium and left ventricle. The left side of the heart is thicker because it pumps blood to the body. The atrium and ventricle on each side are separated by an atrioventricular valve that controls blood flow. The left and right sides of the heart are divided by a septum.

The dog's heart beats about 80 to 130 times per minute. The heart rate, also called the pulse rate, can be felt quite easily on the inside of the upper thigh (at the femoral artery). The palpable pulse is caused by dilatation of the arteries when the expelled blood flows through them. The muscles of the heart ventricles contract during systole and push the blood into the arteries of the body and lungs. The atrioventricular valves close, thus preventing a reflux of blood back into the atrium. Between heart beats the muscles relax. This phase is called diastole and lasts about as long the contraction phase. During diastole, blood flows from the filled atrium into the ventricle.

The heart and pulse rates are usually identical. Dogs with heart disease, however, can have a weak systole that does not cause a pulse wave. Unfortunately many dog breeds are prone to cardiac diseases that often involve the atrioventricular valves. The valves do not close properly, thereby reducing the output of blood during cardiac contractions. Some of the blood leaks back into the atrium. Most affected dogs tend to be from small to medium-sized breeds and of advanced age.

Another heart disease, which is often congenital, affects mainly larger breeds such as Boxers, Newfoundlands, German Shepherds, Rottweilers and Golden Retrievers. The semilunar aortic valve is narrowed, thus reducing the output of blood from the left ventricle.

The heart valves – the atrioventricular valves and semilunar valves – coordinate the blood flow and its direction. (graphic: Maehler)

Diastole
Blood flows into the heart (the heart relaxes))

Systole
Blood is pushed out of the heart (the heart contracts)

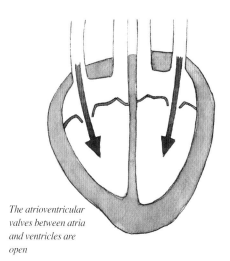

The atrioventricular valves between atria and ventricles are open

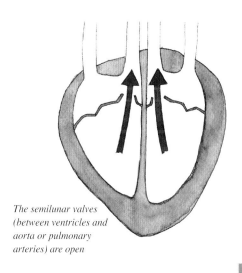

The semilunar valves (between ventricles and aorta or pulmonary arteries) are open

Your vet can listen to the noise of the valves closing with a stethoscope.

The heart has its own nervous system, which is essential so that it can continue to beat in cases of, for example, unconsciousness.

If you notice the following symptoms in your dog you should have a vet check it for a potential cardiac condition: cough, lack of performance, blue tongue or membranes (emergency!), rapid breathing or breathing difficulties, loss of consciousness (emergency!), or loss of appetite without any other obvious reasons.

Circulation

The circulation provides all parts of the body with blood, nutrients, water, hormones, immune cells and oxygen, as well as transporting waste products. The circulation can be divided into three systems: the body part (the largest), the lung part and the liver part. Circulation in the body provides the cells with oxygen and transports blood containing little oxygen (together with carbon dioxide and waste products) back to the heart. The blood is pumped from the left ventricle into the aorta and from there into the arteries of the body down to the capillaries (the smallest blood vessels, which enable the oxygen exchange), before the veins, which join to form the vena cava, transport the blood back towards the heart.

The lung circulation transports the deoxygenated blood to the lungs where it is refreshed with oxygen and carried back to the heart.

The liver circulation collects the blood from the intestines and moves it to the liver, where it is detoxified.

Blood vessels are divided into arteries (which carry blood away from the heart) and veins (which carry blood towards the heart). Arteries can actively contract, using muscles in the arterial walls, thus enabling the oxygenated blood to reach all parts of the body. Veins are generally larger and have less well-developed muscular walls, and they cannot contract actively. Transport back to the heart is helped by the suction force of the heart and also by the muscle pumps (contractions of the skeletal muscles during movement that exert pressure on the veins). Valves in the veins prevent the blood from flowing backwards. Capillaries are the smallest blood vessels and connect arteries and veins. Oxygen exchange between blood and the tissues takes place in the capillaries.

A dog has about 70 millilitres of blood per kilogram of body weight. Therefore, a 20 kilogram dog has a blood volume of roughly one and a half litres.

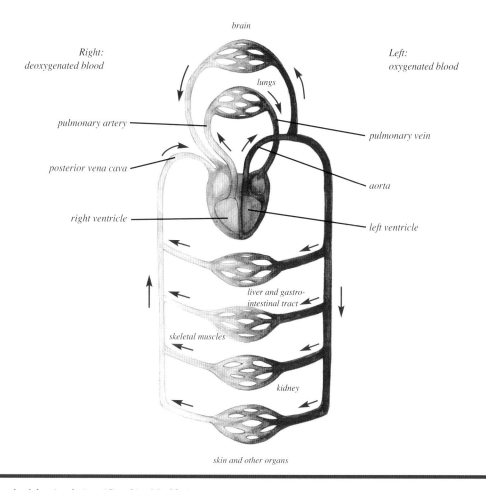

brain

Right:
deoxygenated blood

Left:
oxygenated blood

lungs

pulmonary artery

pulmonary vein

posterior vena cava

aorta

right ventricle

left ventricle

liver and gastro-
intestinal tract

skeletal muscles

kidney

skin and other organs

Sketch of the circulation. (Graphic: Maehler)

Blood is a body fluid that contains diluted substances and cells. The cells are red and white blood cells and platelets (thrombocytes). The platelets regulate blood clotting. Haemoglobin, the red pigment of the blood, is an important part of the red blood cells. It contains iron, which gives the blood its red colour. The main function of haemoglobin is to bind the oxygen in the lungs and aid its transport into the smallest blood vessels. On the way back to the heart haemoglobin also binds the waste product carbon dioxide and transports it to the lungs.

Dogs have different blood types but they differ from human ones. Currently thirteen blood groups are known that are named DEA (dog erythrocyte antigens), with successive numbers. During a first blood transfusion dogs have no antibodies against other blood types. However, they can develop antibodies subsequently and it is important to pay attention to blood types should a second transfusion be required.

(Photo: Lehari)

The lymphatic system

Another important system of the body is the lymphatic system. It plays a vital role in the immune defence of the body – you could picture it as the body's health police.

The lymphatic system is a drainage system for the body's tissues. The tissue fluid,

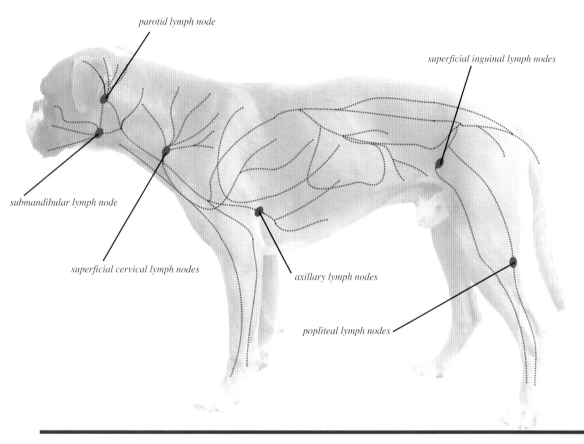

parotid lymph node

superficial inguinal lymph nodes

submandibular lymph node

superficial cervical lymph nodes

axillary lymph nodes

popliteal lymph nodes

Palpable superficial lymph nodes. (Photo and graphic: Mielke)

or lymph, is a bright, clear colour because it contains predominantly water. In the intestinal area it is slightly milky because of a higher high fat content.

Other constituents of lymph fluid are diluted proteins, fat, salt and white blood cells. The lymph is transported via a network of fine lymph vessels to the vena cava and thus back into the bloodstream. The smallest lymph vessels originate from the tissue and are called lymph capillaries. They have no valves and the fluid is transported with the help of body movements. The larger lymph vessels are called lymph ducts. These have valves and are actively involved in the transport of the fluid owing to a muscular layer in their walls. They open into the lymph nodes.

Lymph nodes are important filtration stations that eliminate harmful substances. Several lymph ducts meet at one lymph node but only one duct leaves the node. Each lymph node is responsible for a certain part of the body. The dog has numerous lymph nodes. The lymph ducts that leave the lymph nodes eventually merge and transport the lymph fluid into the vena cava. Lymph vessels therefore play an essential role in the fluid balance of the tissue, in addition to immune defence.

Given that the lymph fluid relies on body movement in order to be transported to the larger lymph vessels, drainage problems often occur in paralysed or inactive dogs. Fluid can build up in the tissues causing visible and palpable swellings – so-called oedema. These swellings can be reduced through external manipulation and massage.

If a lymph node is swollen it indicates a problem in the relevant area. These lymph nodes are working hard to eliminate disease-causing agents and therefore they grow in size.

Lymph nodes can also filter out and destroy particles such as bacteria, foreign bodies and tissue debris. The process of destruction is called cellular phagocytosis, from the Greek word 'phago' meaning 'eating'.

Lymph nodes, as well as the bone marrow, spleen and thymus (a gland present in juvenile dogs that disappears with age), produce several types of lymphocytes. Lymphocytes are a type of leucocyte (white blood cell). Lymphocytes are specific immune cells and make up about 30 per cent of all the white blood cells of a dog.

(Photo: Lehari)

Skin and hair

The skin consists of three layers: the epidermis, dermis and subcutis. The outer skin, the epidermis, is a relatively loose layer that covers the dense and stronger dermis. The epidermis does not contain blood vessels but draws its nutrients from the dermis and subcutis. The dermis and subcutis supply the skin with blood vessels and nerves. The dermis also contains many glands, mainly oil (sebaceous) glands. The oil glands supply the skin with nutrients and protect the skin. The paws contain many sweat glands. Secretions from glands in the ear ensure that dirt is bound and eliminated.

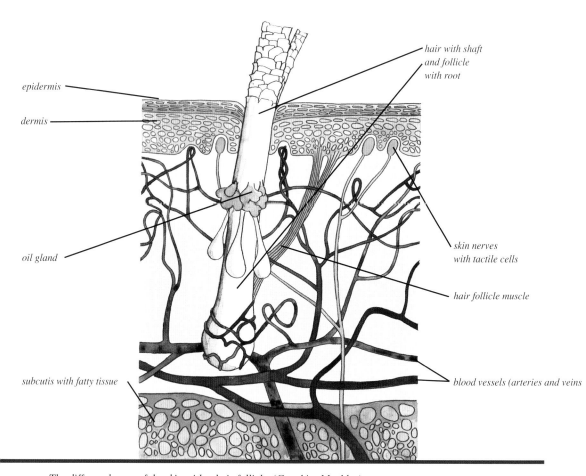

epidermis

dermis

hair with shaft
and follicle
with root

oil gland

skin nerves
with tactile cells

hair follicle muscle

subcutis with fatty tissue

blood vessels (arteries and veins

The different layers of the skin with a hair follicle. (Graphic: Maehler)

Skin and hair are protected by the secretion of the oil glands. This protective product prevents dirt and moisture from damaging the skin and also makes the coat shiny. If a dog is washed with shampoo this protective layer will wash off but cleaning the coat with just water will not destroy the layer. The skin will produce excessive oil after a bath in order to restore the protective layer. If a dog is bathed frequently with shampoo the skin will not be able to maintain a balanced level of protection and excess oil will attract dirt in the coat and cause it to smell strongly. The dog therefore gets dirty again quickly and will smell more intensively than before the wash.

Each individual hair grows out of a hair follicle. A central hair can be surrounded by secondary hairs that also have a hair follicle and glands. Muscles in the follicle allow the hair to be raised (especially the hair of neck and back).

Claws and pads are covered with so-called keratin cells or horn. The horn of the claw is hard whereas pads are made from soft horn. The pads are less heat sensitive than other parts of the skin. The subcutis of the pads contains a layer of fatty cells that act as shock absorbers. Because not all dogs chew their own claws they often have to be trimmed back if they are not worn off sufficiently (e.g. in dogs that are exercised mainly on soft ground). If the claws get too long there is a risk of them growing into the pads. When trimming it is important not to cut into the 'live' area of the claws where the blood vessels are (the quick). The blood vessels are easy to see in light-coloured claws but not in black ones. If in doubt a vet should trim the claws.

Paw of the front leg. (Photo: Mielke)

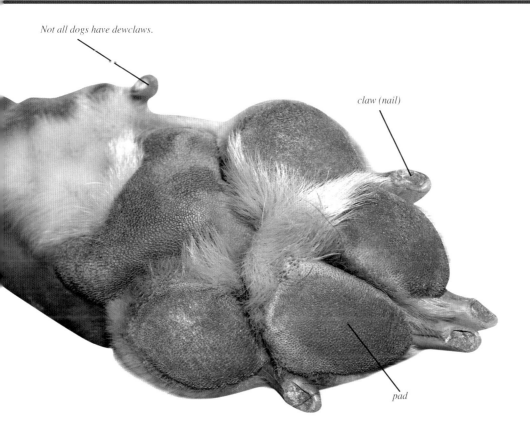

Not all dogs have dewclaws.

claw (nail)

pad

When trimming claws you should not cut them at 90 degrees but maintain the natural angle already shaped from wear.

Skin and hair fulfil many functions: protection against heat and cold, a shield against mechanical forces, chemicals and insects, as well as tactile senses and communication. The transition from wolf to domestic dog has changed not only the body shape but also the hair structure. German Shepherds, for example, have 'Stockhaar' (a smooth coat), Dachshunds are often rough coated, Boxers are smooth and shortcoated, Collies have a long coat, Irish Setters have medium silky coats, Komondors corded coats and Poodles have woolly, curly coats. Moulting is controlled genetically and depends on temperature and daylight. Dogs kept predominantly indoors are subjected to constant temperatures and daylight all year round and may moult permanently. Some breeds such as Poodles do not moult at all; however, they need to be clipped regularly because their hair grows at a constant rate.

The hair can be divided into guard hairs and undercoat. Short-haired breeds such as Boxers lack the undercoat, which leaves them more sensitive to exposure to cold than other, double-coated dogs.

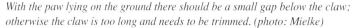

With the paw lying on the ground there should be a small gap below the claw; otherwise the claw is too long and needs to be trimmed. (photo: Mielke)

The
nervous system

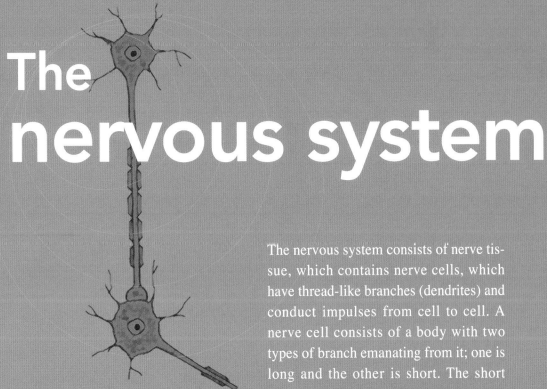

The nervous system consists of nerve tissue, which contains nerve cells, which have thread-like branches (dendrites) and conduct impulses from cell to cell. A nerve cell consists of a body with two types of branch emanating from it; one is long and the other is short. The short

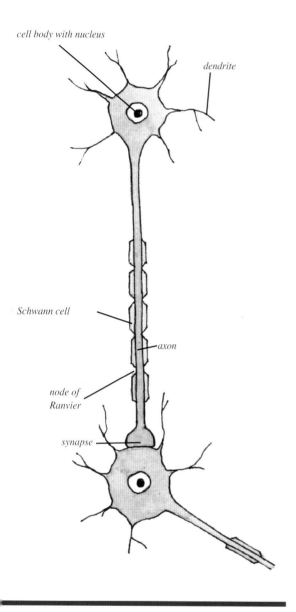

cell body with nucleus

dendrite

Schwann cell

axon

node of Ranvier

synapse

A nerve cell has two types of branch.
(Graphic: Maehler)

gland cell. The end of the axon is called the synapse. The synapse is a control centre that allows an impulse to pass in one direction, like a one-way system. In order to speed up the transfer of impulses, the long branch is surrounded by so-called Schwann cells, interrupted by the ring-like nodes of Ranvier that form gaps between the Schwann cells. The nodes facilitate the transfer of the impulse because the impulse jumps from node to node along the axon.

The nervous system regulates the functions of the internal organs, receives stimuli from the environment through the skin and senses, and sends these to the central nervous system. From there a command is sent back to the muscles. This pathway is sometimes called a reflex.

The nervous system can be divided into three parts: the central nervous system, the autonomic nervous system and the peripheral nervous system.

Central nervous system

The brain and spinal cord belong to the central nervous system. The spinal cord lies within the spinal column. A spinal nerve exits through the gap between adjacent vertebrae. Towards the head, the spinal cord merges with the brain. The spinal cord ends at the sixth lumbar vertebra, where more spinal nerves exit. This area is called the cauda equina because the exiting nerves resemble a horse's tail.

branch receives the impulse from another nerve cell; the long branch (the axon) transmits it to another nerve, or a muscle or

Dogs can suffer from compression of the spinal nerves in this area, which results in so-called cauda equina syndrome.

The central nervous system regulates the functions of the body and controls the skeletal muscles.

Autonomic nervous system

The autonomic nervous system regulates the autonomic, subconscious functions of the body. It therefore controls the heart, circulation, breathing, digestion and metabolism, as well as sexual functions. It is divided into the sympathetic and parasympathetic systems.

The sympathetic ('fight or flight') system includes a chain of nerve cell bodies along both sides of the spine from head to tail. The nerves form long bundles. Nerves leaving the brainstem area of the bundle innervate the head and organs of the chest and abdominal cavities. Other nerves exit in the sacral area and innervate the intestines, bladder and reproductive organs.

The parasympathetic nervous system ensures that all the organs of the gastrointestinal tract function autonomically.

Peripheral nervous system

The peripheral nervous system includes all nerves connecting the spinal cord and brain to the extremities and organs. Nerves that transmit impulses from the brain to the muscles are called motor nerves; the ones that send signals from the senses to the brain are called sensory nerves. Many nerves in the brain and all spinal nerves are mixed nerves, i.e. they consist of both motor and sensory components.

These nerves invade the whole body. You can compare them to high-voltage cables: a thick main cable divides into smaller cables that divide further and supply every house with electricity.

A cross-section through the spinal cord shows a grey, butterfly-shaped inner zone and a white outer zone. (Graphic: Maehler)

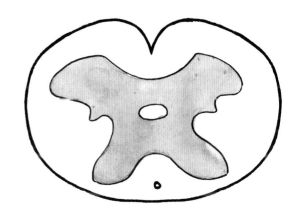

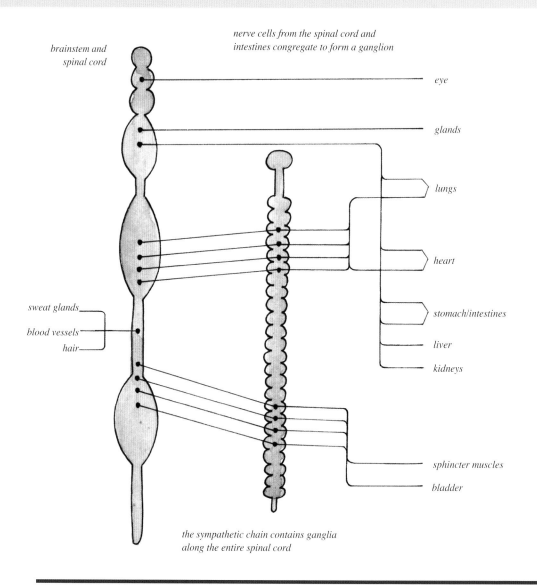

brainstem and spinal cord

nerve cells from the spinal cord and intestines congregate to form a ganglion

eye

glands

lungs

heart

sweat glands

blood vessels

hair

stomach/intestines

liver

kidneys

sphincter muscles

bladder

the sympathetic chain contains ganglia along the entire spinal cord

Overview of the autonomic nervous system. (Graphic: Maehler)

All parts of the nervous system are connected to each other. Each muscle is usually innervated by only a single nerve.

The most important nerves and their motor functions are shown in the appendix.

The senses

A dog has five different senses with their relevant sensory organs: touch (skin and receptors of other organs), taste (taste buds in the tongue), smell (nose), vision (eyes) and hearing (ears). The brain processes all incoming information. The centre for smell in the dog's brain is particularly well developed, which means that the dog's sense of smell is extremely acute. A dog explores its environment by smell first.

Smell

The nose extends from the stop to the tip of the nose, including the nostrils. The area around the nostrils is called the rhinarium. The hairless area is hairless and has a groove in the middle (philtrum). The nose is a vital organ for breathing as well as for smelling. The inhaled air is warmed and moistened inside the nose. Small hairs filter out any particles in the air.

It is commonly said that if a dog has a cold wet nose it is healthy, but this is not always true. There are various reasons that a healthy dog can have a dry or warm nose. Moistening of the nose – or to be precise, of the rhinarium – is provided by a secretion from glands and also tears. The glands are located to one side on the upper jaw and open into the nasal cavity, as does the tear duct.

So why is a dog's nose wet? A wet nose enables the dog to inhale a larger number

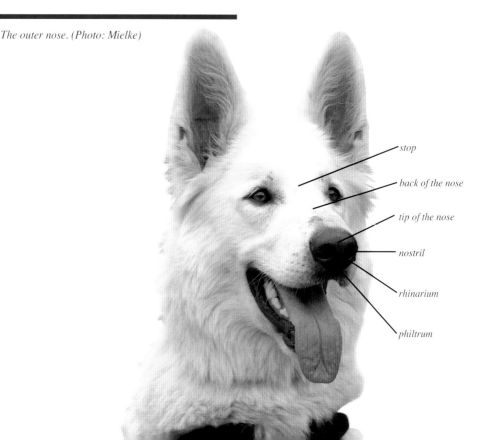

The outer nose. (Photo: Mielke)

stop

back of the nose

tip of the nose

nostril

rhinarium

philtrum

of odour molecules and to transport them to the olfactory region of the nasal membranes. The olfactory epithelium covers the respiratory passages of the nose, also called ethmoturbinate bones. These bones are folded to form an intricate maze of bony lamellae in order to provide a large surface area for the analysis of scent molecules. Many olfactory receptors are located in the membrane. They transform a scent molecule to a chemical message and send signals to the olfactory lobe in the brain.

> The olfactory lobe of the brain has to process countless signals sent from the dog's nose, so trailing work is a mentally demanding and stimulating exercise for the dog.

Olfactory cells are constantly renewed, which is unusual because nerve cells cannot normally be replaced. A dog's sense of smell is at least fifteen times more sensitive than that of humans and they are able to detect odour molecules in very low quantities. Dogs have an additional olfactory scent organ, the so-called vomeronasal or Jacobson's organ, after the scientist who discovered it. The Jacobson's organ is also present in humans and in horses – the latter stimulate it by performing flehmen.

This scent organ has no direct connection to the olfactory membranes of the nose. The scent molecules reach the vomeronasal organ via the nose as well as via a small duct located in the mouth. However, the organ reacts more to the heavy odour molecules from the mouth than to the volatile ones inhaled from the air. In particular, pheromones excreted in body fluids by other dogs are detected by this organ and it is interesting to note that only pheromones from other members of the same species are detected. Pheromones contain information about sex, reproductive status and hierarchy.

Hearing

Wolves and many dogs have upright ears that can be turned individually towards the direction of any sound. Upright ears act like funnels, detecting and directing sound waves. However, even dogs with dropped ears have far better hearing than humans. Dogs can hear much higher frequencies than humans, which is why special dog whistles using high frequency ranges cannot be heard by us. A dog is also able to detect sound from a greater distance than we can.

The ear is also an essential organ for balance and spatial orientation; both faculties are located in the inner ear. The ear consists of three regions: the outer, middle and inner ear. The outer ear consists of the ear lobe and ear

Different shapes of dogs' ears. The upright ears of, for example, the German Shepherd are the most wolf-like. (Photos: Lehari)

canal; it detects and transmits the sound waves. The ear lobe is a cartilaginous structure from which the ear canal leads into the inner parts of the ear. The eardrum forms the border of the middle ear and spans the bony ear canal. Inside the middle ear lie three ear bones or ossicles; they are also called the hammer, anvil and stirrup. They amplify the incoming sound waves but also protect the inner ear from excessive vibration. The inner ear is divided into the cochlea (for hearing) and the vestibular apparatus (for balance). The cochlea transforms the sound waves into chemical signals that are then processed in the brain.

Vision

It was thought for a long time that dogs could only see in black and white with varying shades of grey. Recent research has shown, however, that dogs can also perceive certain colours. Unlike most humans, dogs only possess two, instead of three,

Structure of a dog's ear. (Graphic: Maehler)

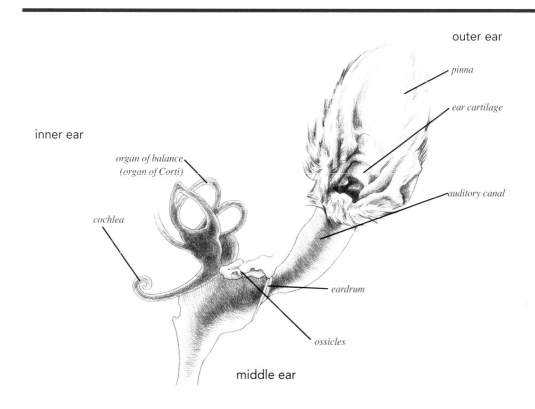

outer ear

pinna

ear cartilage

inner ear

organ of balance (organ of Corti)

auditory canal

cochlea

eardrum

ossicles

middle ear

colour receptors and are therefore red–green colour blind. It is difficult for a dog to see a red ball in green grass. Of course, their acute sense of smell still makes finding the ball quite easy. A blue or yellow ball is found more quickly though, because the eyes and nose can work together. However, if you want to train the dog's sense of smell it is better to use a red ball on a green surface.

Dogs see moving objects better than stationary ones because their eyes are sensitive to movement and light but their visual acuity is inferior to a human's. Their visual field is wider than that of humans owing to the lateral positioning of the eyes, especially in breeds with long muzzles. The visual field is the area in which the eye can detect movement or objects without moving the eye ball. A human visual field encompasses about 190 to 200 degrees, whereas an average dog has a visual field of about 240 degrees, depending on the breed and the position of the eyes. Dogs with short muzzles have a smaller field because their eye positions are more central.

Three-dimensional vision, however, is better in short-muzzled dogs because the visual fields of the individual eyes overlap more than in dogs with more lateral eye positions.

The eye consists of the following parts: the eyelid, cornea, iris, pupil, lens, chamber and retina. The structure and function of an eye can be compared to a camera. The eyelid acts as a lens cap and protects the eye. The cornea shapes the eyeball and can be compared to the camera housing. The iris acts as a lens aperture. It is located in front of the lens between the anterior and posterior chamber and has a small opening, the pupil. The iris contains pigments that express the genetically determined eye colour. Focusing is regulated by the lens

Top: this is how we see the toy.
Bottom: this is how a dog sees it. (Photo: Mielke)

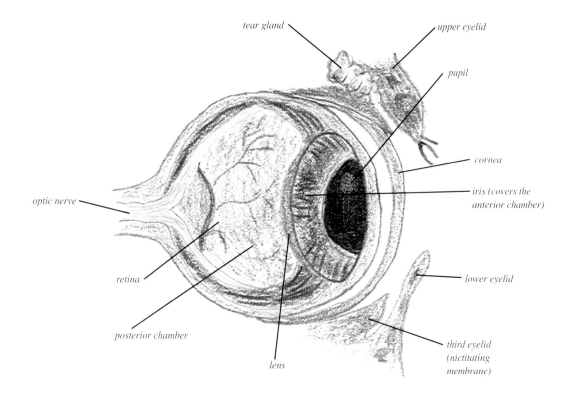

The eye. (Graphic: Mielke)

and its surrounding muscular ciliary body. The retina is made of photoreceptors (the rods and cones) and can be compared to the film or sensor of the camera. The retina sends signals to the brain via the optic nerve. These signals are transformed to pictures in the brain and stored. Dogs, like us, can connect certain experiences and feelings with memorised pictures that are activated each time they appear.

Touch

The sense of touch (somatic sense) is also more developed in dogs than in humans. Dogs possess tactile hairs that can detect the slightest air movements. These hairs are located around the muzzle and above the eyes. Dogs can sense with their whole body, including their paws, owing to sensory nerves that are located everywhere in the body. They process information about touch, pressure, temperature and painful stimuli.

Touching certain skin areas can trigger a reflex reaction. For example, when you stroke a certain part of a dog's abdomen it will begin to scratch itself with its hind leg. The stroking action in this particular area of the skin triggers a reflex that leads to contractions of the hind leg muscles.

Because this reflex occurs at a different location than the triggering stimulus it is a called a polysynaptic reflex. A reflex that affects the same location as the stimulus is a monosynaptic reflex, e.g. the knee cap reflex that occurs in humans as well as dogs: tapping the tendon of the quadriceps muscles just below the knee cap leads to extension of the stifle joint.

This small cross-bred bitch has prominent tactile hairs (whiskers) on the muzzle and above her eyes. (Photo: Mielke)

Taste

Finally, there is something that we can do better than our dogs! Our sense of taste is better developed than that of dogs. Humans possess about six times as many taste buds as dogs. Dogs can still distinguish sweet, sour, bitter and salty tastes. However, because taste and smell work together during eating we can assume that the sense of taste has an inferior role to play in the dog.

Taste buds are located on the surface of the tongue. Nerve cells inside the buds send signals along nerves to the brain where they are analysed. Most taste buds lie in the rear part of the tongue.

Several groups of muscles make up the dog's tongue: the genioglossus, the lateral tongue muscles and the hyoglossus. The tongue muscles enable tongue movements, thus supporting chewing and swallowing the food. A dog can roll its tongue, which facilitates water intake.

taste buds

tip of tongue, can
be rolled inwards

The tongue is a very mobile muscular structure covered by a membrane; most taste buds are located towards the rear portion. (Photo: Mielke)

Appendix

The terms in this chapter are by no means a complete list; the tables merely contain the terms used in the book. They are in alphabetical order of their English as well as Latin names. This allows you to look up the terms you find in the book.

Abbreviations:

M: muscle (musculus)
Mm: muscles
N: nerve (nervus)
Art: joint (articulatio)
Proc: process (processus)
Lig: ligament (ligamentum)

English terms of location

Area	Region
Away from the medial plane of the body	Abduction
Bending a joint	Flexion
Downwards; away from the centre of the body	Distal
Left	Sinister
Overstretching of a joint	Hyperextension
Palm side of the front paw	Palmar
Right	Dexter
Sole side of the hind paw	Plantar
Stretching of a joint	Extension
Towards the abdomen	Ventral
Towards the back	Dorsal
Towards the head	Cranial
Towards the medial plane of the body	Adduction
Towards the middle	Medial
Towards the nasal or oral area	Rostral
Towards the outside	Lateral
Towards the tail	Caudal
Tear	Rupture
Turning	Rotation
Upwards; towards the centre of the body	Proximal

Latin terms of location

Abduction	Away from the medial plane of the body
Adduction	Towards the medial plane of the body
Caudal	Towards the tail
Cranial	Towards the head
Dexter	Right
Distal	Downwards; away from the body
Dorsal	Towards the back
Extension	Stretching of a joint
Flexion	Bending a joint
Hyperextension	Overstretching of a joint
Lateral	Towards the outside
Medial	Towards the middle
Palmar	Palm side of the front paw
Plantar	Sole side of the hind paw
Proximal	Upwards; towards the body
Region	Area
Rostral	Towards the nasal or oral area
Rotation	Turning
Rupture	Tear
Sinister	Left
Ventral	Towards the abdomen

The skeleton – English names

Acromion process	Acromion
Anterior cruciate ligament	Ligamentum cruciatum cranialis
Anterior rim	Margo cranialis
Arch of vertebra	Arcus vertebrae
Arm bone	Humerus
Ball and socket joint	Enarthrosis
Ball joint	Art. sphaeroidea
Body of vertebra	Corpus vertebrae
Bone	Os
Bone development	Ossification
Bone in ankle joint	Talus
Bone marrow cavity	Cavum medullare
Bone material – collagen fibres	Ossein
Bone outgrowth	Osteophyte
Breastbone	Sternum
Collar bone; clavicle	Clavicula
Compact portion of bone	Substantia compacta
Condoyle joint	Art. condylaris
Coracoid bone	Os coracoideus
Crest of shoulder blade	Spina scapulae
Disc nucleus	Nucleus palposa
Elbow eminence	Olecranon
Ellipsoid joint	Art. ellipsoidea
End of a long bone	Epiphysis
False ribs	Costae asternales
Femorotibial joint	Art. femorotibialis
Fibula (calf bone)	Fibula
Flat broad bones	Ossa planae
Foot bones (front)	Ossa metacarpalia
Foot bones (hind)	Ossa metatarsalis
Furrow running down shaft of humerus	Sulcus intertubercularis
Greater trochanter	Trochanter major
Greater tubercle	Tuberculum majus
Head of fibula	Caput fibulae
Head of radius	Caput radii
Heel bone	Calcaneus
Hinge joint	Ginglymus
Hip bone	Os coxae
Hip joint	Art. coxae
Hip joint socket	Acetabulum
Hock bones	Ossa tarsalia
Hole in vertebra	Foramen vertebrae
Ilium	Os ilium
Inner layer of joint capsule	Membrana synovialis
Inner prominence of hock	Malleolus mediales tibiae
Ischial tuberosity	Tuber ischiadicu
Ischium	Os ischii
Joint capsule	Capsula synovia
Joint cartilage	Cartilago articula
Joint condyles	Condylus
Joint fluid	Synovia
Joint ligaments	Ligamenta articu
Joint surface	Facies articularis
Knee cap	Patella
Knee cap joint	Art. femoropatel
Lateral condyle	Condylus lateral
Lateral epicondyle	Epicondylus late
Long bones	Ossa longa
Medial epicondyle	Epicondylus med
Medial joint condyle	Condylus medial
Middle part of a long bone	Diaphysis
Ossification from outside to the inside	Perichondrial ossification
Outer layer of the joint capsule	Membrana fibros
Outer prominence of hock	Malleolus lateral fibulae
Outer portion of bone	Substantia cortic
Pelvic crest	Crista iliaca
Pelvis	Pelvis
Periosteum	Periosteum
Plane gliding joint	Art. plana
Posterior rim	Margo caudalis
Projections inside joint capsule	Villi synoviales
Pubic bone	Os pubis
Pubic symphysis	Symphysis pubis
Radius	Radius
Ribs	Costae
Rim of vertebra	Margo dorsalis
Ring of disc	Annulus fibrosus
Rotary joint	Art. trochoidea
Round ligament of femur head	Ligamentum cap ossis femoris
Sacrum	Os sacrum
Saddle joint	Art. sellaris
Screw joint	Art. cochlearis
Sesamoid bones	Ossa sesamoidea
Accessory carpal bone	Os carpi accesorium
Sheets of bone	Tabulae
Shin bone	Tibia
Short bones	Ossa brevia
Shoulder blade	Scapula

lder blade eminence . . . . Tuberculum
supraglenoidalis
lder joint . . . . Art. humeri
le joint . . . . Art. simplex
ge joint . . . . Art. delabens
al canal . . . . Canalis
intervertebralis
al disc . . . . Discus
intervertebralis
e . . . . Columna vertebralis
al joint . . . . Art. spiralis
gy substance . . . . Substancia spongiosa
e joint . . . . Art. genus
ght patella ligament . . . . Ligamentum patellae

Styloid process . . . . Processus styloideus
Styloid process of the radius . . . . Processus styloideus
radii
Styloid process of the ulna . . . . Processus styloideus
ulnae
Thigh bone . . . . Femur
Toe bones . . . . Phalanges
Trochlea of knee cap . . . . Trochlea patellaris
True joint, synovial joint . . . . Juncturae synovialis
True ribs . . . . Costae sternales
Tuberosity of the tibia . . . . Tuberositas tibiae
Ulna . . . . Ulna
Wrist bones . . . . Ossa carpi

he skeleton – Latin names

abulum . . . . Hip joint socket
mion . . . . Acromion process
lus fibrosus . . . . Ring of disc
s vertebrae . . . . Arch of vertebra
cochlearis . . . . Screw joint
condylaris . . . . Condyle joint
coxae . . . . Hip joint
delabens . . . . Sledge joint
ellipsoidea . . . . Ellipsoid joint
femoropatellaris . . . . Knee cap joint
femorotibialis . . . . Femorotibial joint
genus . . . . Stifle joint
humeri . . . . Shoulder joint
plana . . . . Plane gliding joint
sellaris . . . . Saddle joint
simplex . . . . Simple joint
sphaeroidea . . . . Ball joint
spiralis . . . . Spiral joint
trochoidea . . . . Rotary joint
aneus . . . . Heel bone
lis intervertebralis . . . . Spinal canal
ula synovialis . . . . Joint capsule
t fibulae . . . . Head of fibula
t radii . . . . Head of radius
lago articularis . . . . Joint cartilage
m medullare . . . . Bone marrow cavity
icula . . . . Clavicle (collarbone)
mna vertebralis . . . . Spine
lylus . . . . Joint condyles

Condylus lateralis . . . . Lateral condyle
Condylus medialis . . . . Medial condyle
Corpus vertebrae . . . . Body of vertebra
Costae . . . . Ribs
Costae asternales . . . . False ribs
Costae sternales . . . . True ribs
Crista iliaca . . . . Pelvic crest
Diaphysis . . . . Middle part of a long bone
Discus intervertebralis . . . . Spinal disc
Enarthrosis . . . . Ball and socket joint
Epicondylus lateralis . . . . Lateral epicondyle
Epicondylus medialis . . . . Medial epicondyle
Epiphysis . . . . End of a long bone
Facies articularis . . . . Joint surface
Femur . . . . Thigh bone
Fibula . . . . Fibula (calf bone)
Foramen vertebrae . . . . Hole in vertebra
Ginglymus . . . . Hinge joint
Humerus . . . . Arm bone
Juncturae synovialis . . . . True joint, synovial joint
Ligamenta articularia . . . . Joint ligaments
Ligamentum capitis ossis femoris . . . Round ligament of
femur head
Ligamentum cruciatum cranialis . . . Anterior cruciate ligament
Ligamentum patellae . . . . Straight patellar ligament
Malleolus lateralis fibulae . . . . Outer prominence of hock
Malleolus mediales tibiae . . . . Inner prominence of hock
Margo caudalis . . . . Posterior rim
Margo cranialis . . . . Anterior rim

Margo dorsalis	Rim of vertebra
Membrana fibrosa	Outer layer of the joint capsule
Membrana synovialis	Inner layer of joint capsule
Nucleus palposa	Disc nucleus
Olecranon	Elbow eminence
Os	Bone
Os carpi accessorium	Sesamoid bone
Os coracoideus	Coracoid bone
Os coxae	Hip bone
Os ilium	Ilium
Os ischii	Ischium
Os pubis	Pubic bone
Os sacrum	Sacrum
Ossa brevia	Short bones
Ossa carpi	Wrist bones
Ossa longa	Long bones
Ossa metacarpalia	Front foot bones
Ossa metatarsalis	Hind foot bones
Ossa planae	Flat bones
Ossa sesamoidea	Sesamoid bones
Ossa tarsalia	Hock bones
Ossein	Bone material – collagen fibres
Ossification	Bone development
Osteophyte	Bone outgrowth
Patella	Knee cap
Pelvis	Pelvis
Perichondrial ossification	Ossification from outside to the inside
Periosteum	Periosteum
Phalanges	Toe bones
Processus styloideus	Styloid process
Processus styloideus radii	Styloid process of the
Processus styloideus ulnae	Styloid process of the
Radius	Radius
Scapula	Shoulder blade
Spina scapulae	Crest of shoulder blad
Sternum	Breastbone
Substantia spongiosa	Spongy substance
Substantia compacta	Compact portion of b
Substantia corticalis	Outer portion of bone
Sulcus intertubercularis	Furrow running dowr shaft of humerus
Symphysis pubis	Pubic symphysis
Synovia	Joint fluid
Tabulae	Sheet of bone
Talus	Bone in hock joint
Tibia	Shin bone
Trochanter major	Greater trochanter
Trochlea patellaris	Trochlea of knee cap
Tuber ischiadicum	Ischial tuberosity
Tuberculum majus	Greater tubercle
Tuberculum supraglenoidalis	Shoulder blade emine
Tuberositas tibiae	Tuberosity of the tibia
Ulna	Ulna
Villi synoviales	Projections inside joir capsule

The muscles – English names

Abductor of lower leg	M. abductor cruris caudalis
Adductor muscles	Mm. adductores
Anconeus muscle	M. anconeus
Anterior tibial muscle	M. tibialis cranialis
Arm muscle	M. brachialis
Attachment	Insertio
Biceps muscle	M. biceps brachii
Bursa, fibrous sac filled with synovial fluid	Bursa synovialis
Coracobrachial muscle	M. coracobrachialis
Deep gluteus muscle	M. gluteus profundus
Deep pectoral muscle	M. pectoralis profundus
Deltoid muscle	M. deltoideus
Diaphragm	Diaphragma
Extensor muscles	Extensors
Muscles that bend a joint	Flexors
Gastrocnemius (calf) muscle	M. gastrocnemius
Gracilis muscle	M. gracilis
High tension	Hypertonicity
Iliopsoas muscle	M. iliopsoas
Infraspinatus muscle	M. infraspinatus
Knee joint muscle	M. popliteus
Large abdominal muscle	M. transversus abdom
Large teres muscle	M. teres major
Latissimus dorsi muscle	M. latissimus dorsi
Layer of fibrous tissue	Fascia
Layer of flat, broad tendon	Aponeurosis
Leg–head muscle	M. brachiocephalicus

back muscle	M. longissimus	Semitendinosus muscle	M. semitendinosus
fibular muscle	M. fibularis longus	Sensors that provide information about joint angle, muscle length and tension	Proprioceptors
ension	Hypotonicity	Serratus muscle	M. serratus ventralis
eter (cheek) muscle	M. masseter	Short fibular muscle	M. fibularis brevis
al gluteal muscle	M. gluteus medius	Small teres muscle	M. teres minor
le	Musculus	Splenius capitis muscle	M. splenius capitis
le contraction	Contraction	Stomach	Venter
le wasting	Atrophy	Straight stomach muscle	M. rectus abdominis
les acting against each other	Antagonists	Subscapular muscle	M. subscapularis
les between ribs	Mm. intercostales	Superficial gluteal muscle	M. gluteus superficialis
les involved in exhalation	Expiratory mm.	Superficial pectoral muscle	M. pectoralis superficialis
les involved in inhalation	Inspiratory mm.	Supraspinatus muscle	M. supraspinatus
les working together	Synergists	Temple muscle	M. temporalis
que internal abdominal muscle	M. obliquus internus abdominis	Tendon sheaths	Vaginae synoviales tendinum
que external abdominal muscle	M. obliquus externus abdominis	Tension in the muscle	Muscle tone
transversarius muscle	M. omotransversarius	Tensor of the antebrachial fascia	M. tensor fasciae antebrachii
n	Origio	Tensor of the fascia lata	M. tensor fasciae latae
neus muscle	M. pectineus	Trapezius muscle	M. trapezius
ormis (pear-shaped) muscles	M. piriformis	Triceps muscle	M. triceps brachii
riceps muscle	M. quadriceps femoris	Ulnar carpal extensor muscle	M. extensor carpi ulnaris
al carpal extensor muscle	M. extensor carpi radialis	Ulnar carpal flexor muscle	M. flexor carpi ulnaris
al carpal flexor muscle	M. flexor carpi radialis	Upper thigh muscle	M. biceps femoris
aboid muscle	M. rhomboideus	White line	Linea alba
rius muscle	M. sartorius		
membranosus muscle	M. semimembranosus		

ne muscles – Latin names

uctors	Abductor muscles (away from body)	Hypotonicity	Low tension
uctors	Adductor muscles (towards body)	Insertio	Attachment
gonists	Muscles acting against each other	Inspiratory mm.	Muscles involved in inhalation
neurosis	Layer of flat, broad tendon	Linea alba	White line
phy	Muscle wasting	M. abductor cruris caudalis	Abductor of lower leg
a synovialis	Bursa, fibrous sac filled with synovial fluid	M. anconeus	Anconeus muscle
raction	Muscle contraction	M. biceps brachii	Biceps muscle
hragma	Diaphragm	M. biceps femoris	Upper thigh muscle
ratory mm.	Muscles involved in exhalation	M. brachialis	Arm muscle
nsors	Muscles that extend a joint	M. brachiocephalicus	Leg–head muscle
a	Layer of fibrous tissue	M. coracobrachialis	Coracobrachial muscle
ors	Muscles that bend a joint	M. deltoideus	Deltoid muscle
rtonicity	High tension	M. extensor carpi radialis	Radial carpal extensor muscle
		M. extensor carpi ulnaris	Ulnar carpal extensor muscle
		M. fibularis brevis	Short fibular muscle
		M. fibularis longus	Long fibular muscle
		M. flexor carpi radialis	Radial carpal flexor muscle

M. flexor carpi ulnaris . . . . Ulnar carpal flexor muscle
M. gastrocnemius . . . . Gastrocnemius (calf) muscle
M. gluteus medius . . . . Medial gluteal muscle
M. gluteus profundus . . . . Deep gluteal muscle
M. gluteus superficialis . . . . Superficial gluteal muscle
M. gracilis . . . . Gracilis muscle
M. iliopsoas . . . . Iliopsoas muscle
M. infraspinatus . . . . Infraspinatus muscle
M. latissimus dorsi . . . . Latissimus dorsi muscle
M. longissimus . . . . Long back muscle
M. masseter . . . . Masseter (cheek) muscle
M. obliquus externus abdominis . . . . Oblique external abdominal muscle
M. obliquus internus abdominis . . . . Oblique internal abdominal muscle
M. omotransversarius . . . . Omotransversarius muscle
M. pectineus . . . . Pectineus muscle
M. pectoralis profundus . . . . Deep pectoral muscle
M. pectoralis superficialis . . . . Superficial pectoral muscle
M. piriformis . . . . Piriformis (pear-shaped) muscles
M. popliteus . . . . Knee joint muscle
M. quadriceps femoris . . . . Quadriceps muscle
M. rectus abdominis . . . . Straight abdominal muscle
M. rhomboideus . . . . Rhomboid muscle
M. sartorius . . . . Sartorius muscle

M. semimembranosus . . . . Semimembranosus m
M. semitendinosus . . . . Semitendinosus musc
M. serratus ventralis . . . . Serratus muscle
M. splenius capitis . . . . Splenius capitis musc
M. subscapularis . . . . Subscapularis muscle
M. supraspinatus . . . . Supraspinatus muscle
M. temporalis . . . . Temple muscle
M. tensor fasciae antebrachii . . . . . . Tensor of the antebra fascia
M. tensor fasciae latae . . . . Tensor of the fascia la
M. teres major . . . . Large teres muscle
M. teres minor . . . . Small teres muscle
M. tibialis cranialis . . . . Cranial tibial muscle
M. transversus abdominis . . . . Transverse abdominal
M. trapezius . . . . Trapezius muscle
M. triceps brachii . . . . Triceps muscle
Mm. adductores . . . . Adductor muscles
Mm. intercostales . . . . Muscles between ribs
Muscle tone . . . . Tension in the muscle
Musculus . . . . Muscle
Origio . . . . Origin
Proprioceptors . . . . Sensors that provide i mation about joint an muscle length and ter
Synergists . . . . Muscles that work to
Vaginae synoviales tendinum . . . . . . Tendon sheaths
Venter . . . . Stomach

The organs – English names

Alimentary canal between
mouth and stomach . . . . Oesophagus
Anus . . . . Anus
Bladder . . . . Vesica urinaria
Blood types . . . . Dog erythrocyte antigens 1–13
Blood vessel carrying blood
away from heart . . . . Artery
Blood vessel carrying blood
towards the heart . . . . Vein
Body of nerve cell . . . . Soma
Caecum, part of large intestines . . . . Caecum
Contact area between nerve and
muscle or nerve and nerve . . . . Synapse
Contraction phase of heart . . . . Systole
Deep layer of skin . . . . Subcutis
Digestive gland . . . . Pancreas

End of large intestine . . . . Rectum
Extension of nerve cell . . . . Neurite
External ear . . . . Auris externa
External nose . . . . Nasus externus
Eye . . . . Oculus
Gall bladder . . . . Vesica fella
Heart . . . . Cor
Incorporation of particles
into a cell . . . . Phagocytosis
Inner ear . . . . Auris interna
Inner lining of abdomen . . . . Peritoneum
Inner lining of chest cavity . . . . Pleura
Kidney . . . . Ren
Large intestine . . . . Colon
Large intestine . . . . Intestinum crassum
Layer surrounding heart . . . . Pericardium
Liver . . . . Hepar

end of a nerve cell	Axon	Red blood cells	Erythrocytes
	Pulmo	Red blood pigment	Haemoglobin
airway inside lungs	Bronchus	Relaxation phase of heart	Diastole
body artery	Aorta	Sensitive layer of skin	Dermis
e ear	Auris media	Short end of nerve cell	Dendrite
bundle	Ganglion	Small intestine	Intestinum tenua
skin layer	Epidermis	Smallest blood vessel	Capillary
ympathetic nervous system	Parasympathicus	Soft tissue swelling	Oedema
athetic nervous system	Sympathicus	Special white blood cells	Lymphocytes
f small intestine	Ileum	Stomach	Gaster
f small intestine	Jejunum	Supply from nerve	Innervation
f small intestine	Duodenum	White blood cells	Leucocytes
f throat (voice box)	Larynx	Windpipe	Trachea

he organs – medical names

	Anus	Leucocytes	White blood cells
	Main body artery	Lymphocytes	Special white blood cells
y	Blood vessel carrying blood away from heart	Nasus externus	External nose
externa	External ear	Neurite	Extension of nerve cell
interna	Inner ear	Oculus	Eye
media	Middle ear	Oedema	Soft tissue swelling
	Long end of a nerve cell	Oesophagus	Alimentary canal between mouth and stomach
chus	Main airway inside lungs	Pancreas	Digestive gland
um	Caecum, part of large intestines	Parasympathicus	Parasympathetic part of autonomic nervous system
	Part of large intestine	Peritoneum	Inner lining of abdomen
	Heart	Pericardium	Layer surrounding heart
rythrocyte antigens		Phagocytosis	Incorporation of particles into a cell
) 1-13	Blood types	Pleura	Inner lining of chest cavity
rite	Short end of nerve cell	Pulmo	Lung
is	Sensitive layer of skin	Rectum	End of large intestine
ole	Relaxation phase of heart	Ren	Kidney
enum	Part of small intestine	Soma	Body of nerve cell
rmis	Outer skin layer	Subcutis	Deep layer of skin
rocytes	Red blood cells	Sympathicus	Sympathetic part of autonomic nervous system
lion	Nerve bundle	Synapse	Contact area between nerve and muscle or nerve and nerve
r	Stomach		
oglobin	Red blood pigment		
r	Liver		
	Part of small intestine	Systole	Contraction phase of heart
vation	Supply from nerve	Trachea	Windpipe
inum crassum	Large intestine	Vein	Blood vessel carrying blood towards the heart
inum tenua	Small intestine		
um	Part of small intestine	Vesica fella	Gall bladder
lary	Smallest blood vessel	Vesica urinaria	Bladder
x	Part of throat (voice box)		

Important nerves and the functions of the relevant muscles

Nerve	Functions of the innervated muscles
Musculocutaneus	Extension and adduction of shoulder joint; flexion of elbow joint
Radial	Flexion of shoulder joint; extension of elbow joint, toes and carpal joint; rotation of lower front leg
Caudal gluteal	Extension of hip joint
Cranial gluteal	Flexion of hip joint; extension of stifle joint
Obturator	Adduction and flexion of hip joint; extension of stifle and hock
Femoral	Adduction and flexion of hip joint; extension of stifle joint
Ulnar	Flexion and rotation of lower front leg
Axillary	Flexion of shoulder joint
Sciatic, divided into:	
Fibular	Flexion of hock; extension of toes
Tibial	Flexion of stifle joint; rotation of lower hind leg; extension of hock; flexion of toes

Literature

Should you wish to study canine anatomy more deeply than this book can offer, I recommend the following books for further reading:

R. Micheal Akers, D. Micheal Denbow
Anatomy & Physiology of Domestic Animals
Wiley Blackwell, 2008

Stanley H. Done, Peter C. Goody, Susan A, Evans, Neil C. Stickland
Color Atlas of Veterinary Anatomy, Volume 3, The Dog and Cat
Mosby 2009